CRACKING THE REGENTS
CHEMISTRY

THE PRINCETON REVIEW

1998-99

CRACKING THE REGENTS

CHEMISTRY

NILANJEN SEN

1998–99 Edition
Random House, Inc.
New York, 1998
www.randomhouse.com

Princeton Review Publishing
2315 Broadway, 3rd Floor
New York, NY 10024
E-mail: info@review.com

Published in the United States by Random House, Inc., New York, and
simultaneously in Canada by Random House of Canada Limited, Toronto.

ISBN 0-375-75072-X
ISSN 1097-1114

Editor: Rachel Warren
Design & Production: Meher Khambata & Greta Englert
Production Editor: JoAnn Schambier, Silverchair Science + Communications

9   8   7   6   5   4   3   2   1

1998–99 Edition

ACKNOWLEDGMENTS

This book is dedicated to my parents, Subrata and Tuhina Sen, and my sister, Sudakshina Sen. I would like to express my gratitude to the publishing department at The Princeton Review and to my editor, Rachel Warren, for her assistance and guidance. I also thank the reviewers of this book, Sasha Alcott and Barry Zimmerman. A special thank you goes to Kim Magloire and Paul Maniscalco. Last but not least, I would like to express my sincere gratitude to my former teachers and the hundreds of students who inspired me to pursue a career in education.

CONTENTS

PART I

HOW TO CRACK
THE SYSTEM

INTRODUCTION

This book is for students who want to raise their scores on the Regents examination in chemistry. At the Princeton Review we know what it takes to do well on standardized tests. So work with us, and we can help you raise your score.

The Regents examination in chemistry is composed of 116 questions. Part 1 consists of 56 multiple-choice questions and is worth a total of 65 points. Part 2 is made up of twelve groups, each of which contains five questions; you are expected to choose seven of the groups and complete all of the questions in them. The total point value of Part 2 is 35. The highest possible score on the test is 100 points, but you must score a minimum of 65 points to pass.

WHAT IS THE KEY TO DOING WELL ON THE REGENTS?

There are many things you can do to prepare for the Regents exam. First, remember that this is a standardized test, which means that all you have to do is pick the right answer. No one cares how you got the answer as long as it is the correct answer, so the trick is to learn to recognize the right answer even if you are not sure why it is right. Hey, you have 1 out of 4 chances to choose the right answer!

Studying the right material is the key to doing well on the test. At The Princeton Review, our strategy involves knowing what types of chemistry concepts are almost certain to show up on the test and in what form they will show up. In other words, we have identified the high-frequency questions.

Along with studying the material, an important preparation for the test is to formulate an aggressive strategy for your study. For instance, highlight the key concepts you need to know instead of just reading your text with little or no energy.

HOW MUCH CHEMISTRY DO YOU NEED TO KNOW?

We know exactly what material is tested on the Regents exams and how it is presented. The exams are based on the New York State Regents Chemistry Syllabus.

As mentioned before, Part 1 of the exam consists of 56 multiple-choice questions worth a total of 65 points. These questions test concepts from nine different units taken from the chemistry syllabus:

1. Matter and Energy
2. Atomic Structure
3. Bonding
4. Periodic Table
5. Mathematics of Chemistry
6. Kinetics and Equilibrium
7. Acids and Bases
8. Redox Reactions and Electrochemistry
9. Organic Chemistry

You should focus on these nine units in your prepartion for Part 1 of the Regents exam.

Part 2 of the Regents consists of 60 multiple-choice questions that are divided into twelve groups of five questions each. You must choose seven of the twelve groups and answer them in their entirety. The maximum number of points that can be earned in this section is 35. Part 2 covers the following topics:

1. Matter and Energy
2. Atomic Structure
3. Bonding
4. Periodic Table
5. Mathematics of Chemistry
6. Kinetics and Equilibrium
7. Acids and Bases
8. Redox and Electrochemistry
9. Organic Chemistry
10. Applications of Chemical Principles
11. Nuclear Chemistry
12. Laboratory Activities

This may seem like a lot of information, but for each topic there are only a few key facts you need to know. The chemistry textbook you use in your class goes into far greater detail about each topic than you need to know for the Regents examination.

CRACKING THE REGENTS

There are five strategies that you should use when you take the test:

1. Efficient use of time
2. The three-pass (easy-to-hard) system
3. Process of elimination
4. Intellectual guessing
5. Mnemonics

Strategy 1: Efficient Use of Time

One of the main reasons that taking the Regents is so stressful is that you have only approximately 2 minutes to answer each question, and although 2 minutes might seem like a long time, under testing conditions 2 minutes can go by very fast. If you had all day to do it, you would probably get a much higher score on the test, but when you rush, you are far more likely to make careless mistakes, misread questions, and generally fall into traps. The key to doing well is to pace yourself; know exactly how much time you can afford to spend solving each question.

Examine the entire test before you start responding to the questions. Remember that questions in Part 1 of the test are worth more points than the questions in Part 2. Doesn't it make sense to spend more time solving questions that are worth more? Prepare a schedule for yourself. Allow yourself more time to solve Part 1 questions. Because you have to answer all 56 questions in Part 1, it is to your advantage to do as well as you can in this section. Remember that you get to choose the questions you want to answer in Part 2. So if you are stuck on one group of questions in Part 2, you can always move on and select another group.

One of the best ways to efficiently use your time is to answer the easier questions first. If you are stuck on a hard question, you shouldn't be afraid to skip it so that you can answer an easier question first. The goal is to correctly answer as many questions as you can as quickly as possible. Once you have answered the easier questions, you can spend the remaining time on questions that initially stumped you.

Strategy 2: The Three-Pass (Easy-to-Hard) System

As we've said, the best way to rack up points on this test is to focus on the easiest questions first. Many of the questions are straightforward and should require little effort on your part. You will know when you see a straightforward question because the answer will jump at you when you look through the answer choices, so just nail it and move on. As you read each question, immediately assign it a level of difficulty: easy, medium, or hard. During the first pass, answer all the questions that you have designated as easy. When you come across a hard question, remind yourself that easy questions are worth just as much as hard questions and move on in search of those that are less time-consuming.

During the second pass, answer all of the medium-difficulty questions. These questions are designated medium because they are somewhat time-consuming. More specifically, (1) the question requires you to think about more than one scientific principle before you can formulate an answer, or (2) the answers have to be looked at closely and analyzed before you can settle on a choice. When you come across a question like this, you must slow down so that you can read the question more deliberately and understand what the test makers want you to answer.

If you come across a question that seems very difficult, save it for the third pass.

Strategy 3: Process of Elimination

On most class tests, you need to know your material backward and forward to calculate or determine the correct answers. In other words, if you did not study and had no prior knowledge of the answer, it is likely that you won't be able to answer the questions. This is not the case on the Regents exam. You can do well on this test without knowing the right answer if you can identify all the wrong answers. Let's take a look at an example:

According to Reference Table L, which substance is amphoteric?

(1) HNO_3 (3) HCO_3^-

(2) NO_2^- (4) HF

Does it matter if you recognize any one of these chemical species? Does it matter if you can name these chemical compounds? Absolutely not. To correctly answer this question, you need to have a minimum knowledge of what an amphoteric substance is. If you know that an amphoteric substance can act as both an acid and a base, you know that any substance that acts as an acid must bear a hydrogen (H). And any substance that can act as a base must have electrons to donate (indicated by a negative sign). Therefore, an amphoteric substance must have a hydrogen in order to act as an acid, and a negative charge in order to act as a base.

So try to eliminate the answer choices until you are left with one. Choices 1 and 4 have hydrogen atoms as part of their structure, but they are not negatively charged. Therefore, they cannot be the right answer. Choice 2 has a negative charge but no hydrogen. This choice is not the correct answer either. So you are left with choice 3. Whether you know it is the correct answer is no longer an issue. You are left with one choice because you have concluded that the other three choices can't be the right answer.

Strategy 4: Intellectual Guessing

Because there is no penalty for getting a question wrong, it is to your advantage to guess the answer if you are stuck on a question. Answer all of the questions you know. But when it is time to guess, guess wisely. If you guess on five questions, odds are you will get at least one right. That's one extra point that you wouldn't have if you left all five answers blank.

If you can eliminate a couple of answer choices, your odds of answering the question correctly, even if you guess, are greatly increased. The trick is to intellectually eliminate two of the four choices; then you have a 50 percent chance of guessing the answer correctly. One thing you can do to help yourself is to simplify a complicated question by rephrasing it into simple English. For example, a question worded in the following manner may be difficult to comprehend because it is so long:

> Compared to the average kinetic energy of 1 mole of water at 0°C, the average kinetic energy of 1 mole of water at 298 K is

This question is much easier to understand if it is rephrased as

> Does 1 mole of water have more kinetic energy at 0°C or 298 K?

Learn how to cross-reference facts between questions. A critical piece of information that you need to answer a question can often be found in another question. Stay alert: Information that you need to do well may be scattered throughout the test. You must draw on your reasoning ability to chain together a series of facts if necessary.

Strategy 5: Mnemonics

One of the keys to simplifying chemistry is to organize scientific terms—names of molecules, reactions, theories, and so forth—into a handful of easily remembered words. The best way to accomplish this is by using mnemonics, a device for remembering words. For example, let's take a look at how you can remember the function of a redox reaction through a mnemonic.

- Mnemonic: LEO says GER.

  Lose Electron = Oxidation

  Gain Electrons = Reduction

Mnemonics can be unusual, as long as they help you to remember. Use mnemonics as much as possible. Be creative. The important thing is that you remember the scientific fact.

REVIEW

Before we get started, let's do a quick review of the strategies.

1. Efficient use of time: Know how much time to spend on each question. Keep track of time.

2. The three-pass (easy-to-hard) system: Focus on answering the easier questions before proceeding to the hard questions.

3. Process of elimination: You don't need to know the right answer to correctly answer the question if you can identify the incorrect answers.

4. Intellectual guessing: Use your reasoning ability to eliminate two of the four answer choices so that you can increase your chances of guessing correctly.

5. Mnemonics: Easy-to-remember words or statements help you to recall important scientific facts and theories.

IMPORTANT CHEMICAL TERMS

Absolute zero   The lowest possible temperature (0 K or –273°C)

Acid   A compound that can donate a proton (H^+)

Acid ionization constant   The equilibrium constant defining the degree of dissociation of an acid

Activated complex   The high-energy intermediate product that is formed when reactants react to form products

Activation energy   The minimum energy needed to start a reaction

Alkali metals   The elements in group I in the periodic table

Alkaline metals   The elements in group II in the periodic table

Alkane   Organic molecules that are completely composed of single bonds

Alkene   Organic molecules that are composed of at least one double bond

Alkyne   Organic molecules that are composed of at least one triple bond

Anion   An atom or molecule with a negative charge

Anode   The negatively charged electrode in which oxidation occurs

Aqueous   A type of solution in which water is the solvent

Atom   The smallest unit of an element

Atomic number   The number of protons in the nucleus of an element

Atomic weight   The weight, in grams, of one mole of an element

Avogadro's number 6.02×10^{23}, the number of molecules in one mole of a substance

Base   A compound that releases OH^- ions in solution

Beta (β) particle   An electron emitted from a nucleus during radioactivity

Boyle's law   The law that states that the volume of gases is inversely proportional to the pressure

Catalyst   A substance or a molecule that speeds up a chemical reaction without acting as a reactant

Cathode   The positive electrode at which reduction occurs

Charles' law   The law that states that the volume of gases varies directly with temperature

Compound   A substance formed by the union of two or more elements

Concentration   The relative amount of a solute in a solution

Covalent bond   Nonmetals bonded together by sharing valence electrons

Critical point   A point in a phase diagram where liquid and gas states cease to be distinct

Dissociation   The breakdown of a solute into its constituent ions

Electrolysis   The decomposition of substances by the use of electric current

Electrolyte   An ionic compound that has high electrical conductivity

Electron   A negatively charged subatomic particle

Electronegativity   A number that describes the attraction of an element for electrons in a chemical bond

Endothermic   A type of reaction in which heat is consumed to form products

Exothermic   A type of reaction in which heat is released when products are formed

Equilibrium constant   The ratio of concentrations of products to reactants for a reaction

Free energy   The thermodynamic quantity measuring the tendency of a reaction to proceed

Freezing point   The temperature at which a liquid changes to a solid

Fusion   A melting process

Gram formula weight   An amount of a substance equal in grams to the sum of the atomic weights

Ground state   The electron configuration of lowest energy of an atom

Group   A column of elements in the periodic table

Half reaction   An oxidation or reduction reaction that takes place as part of a redox reaction

Hydrocarbon   An organic compound containing only carbon and hydrogen

Hydroxyl   The OH^- ion

Inert gases   Also called *noble gases*; gases found in group 18 in the periodic table

Ion   An atom with a charge (negative or positive) due to gain or loss of electrons

Ionization   The process that adds or subtracts electrons from atoms

Ionization energy   The amount of energy needed to remove electrons from atoms

Isomers   Molecules that have the same molecular formula but different molecular structures

Isotopes   The same elements that have different numbers of neutrons in their nuclei

Le Châtelier's principle   The principle that states the conditions that disturb and stabilize chemical systems

Melting point   The temperature at which a solid changes to a liquid

Molality   The number of moles of solute in 1 kg of solvent

Molarity   The number of moles of solute in 1 liter of solution

Molecular formula   The ratio of elements in a molecule

Molecule   A group of atoms bonded to each other

Neutralization   The chemical reaction of a strong acid and a strong base that yields a salt and water

Neutron   A subatomic particle with zero charge found inside the nucleus of an element

Nucleus   The core of an atom that is composed of protons and neutrons

Organic   A compound that contains carbon

Oxidation   The reaction process through which elements lose electrons

Periodic table   A chart that displays all the elements ordered by atomic number

pH   A number describing the concentration of hydrogen ions in a solution

Polyprotic   A type of acid with more than one hydrogen atom that can dissociate in solution

Product   A substance formed in excess in a reaction

Proton   A subatomic particle with a positive charge found inside the nucleus of an atom

Reactant   Substances that combine to form products in chemical reactions

Redox   A type of reaction with simultaneous reduction and oxidation

Reduction   The reaction process through which element(s) gain electrons

Salt   A solid compound composed of a metal and a nonmetal

Shell   A set of electron orbitals that have the same principal quantum number

Solute   The substance that is dissolved in solution

Solvent   The liquid substance in which solutes are dissolved

STP   0°C or 273 K and 1 atmosphere

Sublimation   The transformation of a solid directly into a gas without going through the liquid phase

Temperature   Measurement of the average kinetic energy of all atoms of a substance

Titration   The addition of a known volume of solution to another solution (known or unknown) to determine the concentration of that solution

Transition element   An element with atoms that contain unfilled d sublevels

Transmutation   The nuclear process through which one element is converted to another element

Triple point   A point in a phase diagram where the three states of matter are in equilibrium

Valence electrons   The outermost shell of electrons

REFERENCE TABLES

Reference Table A

Physical Constants and Conversion Factors

Name	Symbol	Value(s)	Units
Angstrom unit	Å	1×10^{-10} m	meter
Avogadro's number	N_A	6.02×10^{23} per mol	
Charge of electron	e	1.60×10^{-19} C	coulomb
Electron volt	eV	1.60×10^{-19} J	joule
Speed of light	c	3.00×10^8 m/s	meters/second
Planck's constant	h	6.63×10^{-34} J•s	joule•second
		1.5×10^{-37} kcal•s	kilocalorie•second
Universal gas constant	R	0.0821 L•atm/mol•K	liter-atmosphere/mole•kelvin
		1.98 cal/mol•K	calories/mole•kelvin
		8.31 J/mol•K	joules/mole•kelvin
Atomic mass unit	μ(amu)	1.66×10^{-24} g	gram
Volume standard, liter	L	1×10^3 cm^3 = 1 dm^3	cubic centimeters, cubic decimeter
Standard pressure, atmosphere	atm	101.3 kPa	kilopascals
		760 mm Hg	millimeters of mercury
		760 torr	torr
Heat equivalent, kilocalorie	kcal	4.18×10^3 J	joules

Physical Constants for H_2O

Molal freezing point depression	1.86°C
Molal boiling point elevation	0.52°C
Heat of fusion	79.72 cal/g
Heat of vaporization	539.4 cal/g

Reference Table B

Standard Units

Symbol	Name	Quantity
m	meter	length
kg	kilogram	mass
Pa	pascal	pressure
K	kelvin	thermodynamic temperature
mol	mole	amount of substance
J	joule	energy, work, quantity of heat
s	second	time
C	coulomb	quantity of electricity
V	volt	electrical potential, potential difference
L	liter	volume

Selected Prefixes		
Factor	Prefix	Symbol
10^6	mega	M
10^3	kilo	k
10^{-1}	deci	d
10^{-2}	centi	c
10^{-3}	milli	m
10^{-6}	micro	μ
10^{-9}	nano	n

Reference Table C

Density and Boiling Points of Some Common Gases

Name		Density g/L at STP*	Boiling Point (at 1 atm) K
Air	—	1.29	—
Ammonia	NH_3	0.771	240
Carbon dioxide	CO_2	1.98	195
Carbon monoxide	CO	1.25	82
Chlorine	Cl_2	3.21	238
Hydrogen	H_2	0.0899	20
Hydrogen chloride	HCl	1.64	188
Hydrogen sulfide	H_2S	1.54	212
Methane	CH_4	0.716	109
Nitrogen	N_2	1.25	77
Nitrogen (II) oxide	NO	1.34	121
Oxygen	O_2	1.43	90
Sulfur dioxide	SO_2	2.92	263

*STP is defined as 273 K and 1 atm.

Reference Table D

Solubility Curves

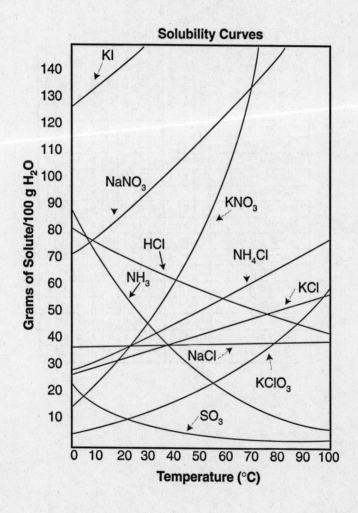

Reference Table E

Table of Solubilities in Water

i—nearly insoluble ss—slightly soluble s—soluble d—decomposes n—not isolated	Acetate	Bromide	Carbonate	Chloride	Chromate	Hydroxide	Iodide	Nitrate	Phosphate	Sulfate	Sulfide
Aluminum	ss	s	n	s	n	i	s	s	i	s	d
Ammonium	s	s	s	s	s	s	s	s	s	s	s
Barium	s	s	i	s	i	s	s	s	i	i	d
Calcium	s	s	i	s	s	ss	s	s	i	ss	d
Copper II	s	s	i	s	i	i	n	s	i	s	i
Iron II	s	s	i	s	n	i	s	s	i	s	i
Iron III	s	s	n	s	i	i	n	s	i	ss	d
Lead	s	ss	i	ss	i	i	ss	s	i	i	i
Magnesium	s	s	i	s	s	i	s	s	i	s	d
Mercury I	ss	i	i	i	ss	n	i	s	i	ss	i
Mercury II	s	ss	i	s	ss	i	i	s	i	d	i
Potassium	s	s	s	s	s	s	s	s	s	s	s
Silver	ss	i	i	i	ss	n	i	s	i	ss	i
Sodium	s	s	s	s	s	s	s	s	s	s	s
Zinc	s	s	i	s	s	i	s	s	i	s	i

Reference Table F

Selected Polyatomic Ions

Hg_2^{2+}	dimercury (I)	CrO_4^{2-}	chromate
NH_4^+	ammonium	$Cr_2O_7^{2-}$	dichromate
$C_2H_3O_2^-$ }	acetate	MnO_4^-	permanganate
CH_3COO^-		MnO_4^{2-}	manganate
CN^-	cyanide	NO_2^-	nitrite
CO_3^{2-}	carbonate	NO_3^-	nitrate
HCO_3^-	hydrogen carbonate	OH^-	hydroxide
		PO_4^{3-}	phosphate
$C_2O_4^{2-}$	oxalate	SCN^-	thiocyanate
ClO^-	hypochlorite	SO_3^{2-}	sulfite
ClO_2^-	chlorite	SO_4^{2-}	sulfate
ClO_3^-	chlorate	HSO_4^-	hydrogen sulfate
ClO_4^-	perchlorate	$S_2O_3^{2-}$	thiosulfate

Reference Table G

Standard Energies of Formation of Compounds at 1 atm and 298 K

Compound	Heat (Enthalpy) of Formation* kcal/mol (ΔHf)	Free Energy of Formation kcal/mol (ΔGf)
Aluminum oxide $Al_2O_3(s)$	−400.5	−378.2
Ammonia $NH_3(g)$	−11.0	−3.9
Barium sulfate $BaSO_4(s)$	−352.1	−325.6
Calcium hydroxide $Ca(OH)_2(s)$	−235.7	−214.8
Carbon dioxide $CO_2(g)$	−94.1	−94.3
Carbon monoxide $CO(g)$	− 26.4	−32.8
Copper (II) sulfate $CuSO_4(s)$	−184.4	−158.2
Ethane $C_2H_6(g)$	−20.2	−7.9
Ethene (ethylene) $C_2H_4(g)$	12.5	16.3
Ethyne (acetylene) $C_2H_2(g)$	54.2	50.0
Hydrogen fluoride $HF(g)$	−64.8	−65.3
Hydrogen iodide $HI(g)$	6.3	0.4
Iodine chloride $ICl(g)$	4.3	−1.3
Lead (II) oxide $PbO(s)$	−51.5	−45.0
Magnesium oxide $MgO(s)$	−143.8	−136.1
Nitrogen (II) oxide $NO(g)$	21.6	20.7
Nitrogen (IV) oxide $NO_2(g)$	7.9	12.3
Potassium chloride $KCl(s)$	−104.4	−97.8
Sodium chloride $NaCl(s)$	−98.3	−91.8
Sulfur dioxide $SO_2(g)$	−70.9	−71.7
Water $H_2O(g)$	−57.8	−54.6
Water $H_2O(\ell)$	−68.3	−56.7

*Minus sign indicates an exothermic reaction.

Sample equations:

$$2Al(s) + \tfrac{3}{2}O_2(g) \rightarrow Al_2O_3(s) + 400.5 \text{ kcal}$$

$$2Al(s) + \tfrac{3}{2}O_2(g) \rightarrow Al_2O_3(s) \quad \Delta H = -400.5 \text{ kcal/mol}$$

Reference Table H

Selected Radioisotopes

Nuclide	Half-Life	Decay Mode
^{198}Au	2.69 d	β^-
^{14}C	5730 y	β^-
^{60}Co	5.26 y	β^-
^{137}Cs	30.23 y	β^-
^{220}Fr	27.5 s	α
^{3}H	12.26 y	β^-
^{131}I	8.07 d	β^-
^{37}K	1.23 s	β^+
^{42}K	12.4 h	β^-
^{85}Kr	10.76 y	β^-
^{85m}Kr*	4.39 h	γ
^{16}N	7.2 s	β^-
^{32}P	14.3 d	β^-
^{239}Pu	2.44×10^4 y	α
^{226}Ra	1600 y	α
^{222}Rn	3.82 d	α
^{90}Sr	28.1 y	β^-
^{99}Tc	2.13×10^5 y	β^-
^{99m}Tc*	6.01 h	γ
^{232}Th	1.4×10^{10} y	α
^{233}U	1.62×10^5 y	α
^{235}U	7.1×10^8 y	α
^{238}U	4.51×10^9 y	α

y = years; d = days; h = hours; s = seconds

*m = metastable or excited state of the same nucleus. Gamma
decay from such a state is called an isomeric transition (IT).
Nuclear isomers are different energy states of the same
nucleus, each having a different measurable lifetime.

Reference Table I

Heats of Reaction at 1 atm and 298 K

Reaction	ΔH (kcal)
$CH_4(g) + 2O_2(g) \longrightarrow CO_2(g) + 2H_2O(\ell)$	−212.8
$C_3H_8(g) + 5O_2(g) \longrightarrow 3CO_2(g) + 4H_2O(\ell)$	−530.6
$CH_3OH(\ell) + \frac{3}{2} O_2(g) \longrightarrow CO_2(g) + 2H_2O(\ell)$	−173.6
$C_6H_{12}O_6(s) + 6O_2(g) \longrightarrow 6CO_2(g) + 6H_2O(\ell)$	−669.9
$CO(g) + \frac{1}{2} O_2(g) \longrightarrow CO_2(g)$	−67.7
$C_8H_{18}(\ell) + \frac{25}{2} O_2(g) \longrightarrow 8CO_2(g) + 9H_2O(\ell)$	−1302.7
$KNO_3(s) \xrightarrow{H_2O} K^+(aq) + NO_3^-(aq)$	+8.3
$NaOH(s) \xrightarrow{H_2O} Na^+(aq) + OH^-(aq)$	−10.6
$NH_4Cl(s) \xrightarrow{H_2O} NH_4^+(aq) + Cl^-(aq)$	+3.5
$NH_4NO_3(s) \xrightarrow{H_2O} NH_4^+(aq) + NO_3^-(aq)$	+6.1
$NaCl(s) \xrightarrow{H_2O} Na^+(aq) + Cl^-(aq)$	+0.9
$KClO_3(s) \xrightarrow{H_2O} K^+(aq) + ClO_3^-(aq)$	+9.9
$LiBr(s) \xrightarrow{H_2O} Li^+(aq) + Br^-(aq)$	−11.7
$H^+(aq) + OH^-(aq) \longrightarrow H_2O(\ell)$	−13.8

Reference Table J

Symbols Used in Nuclear Chemistry

alpha particle	^4_2He	α
beta particle (electron)	$^0_{-1}e$	β^-
gamma radiation		$\sqrt{}$
neutron	^1_0n	n
proton	^1_1H	p
deuteron	^2_1H	
triton	^3_1H	
positron	$^0_{+1}e$	β^+

Reference Table K

Ionization Energies and Electronegativities

1							18
H 313 ← First Ionization Energy (kcal/mol of atoms) 2.2 ← Electronegativity*							He 567
	2	13	14	15	16	17	
Li 125 1.0	Be 215 1.5	B 191 2.0	C 260 2.6	N 336 3.1	O 314 3.5	F 402 4.0	Ne 497
Na 119 0.9	Mg 176 1.2	Al 138 1.5	Si 188 1.9	P 242 2.2	S 239 2.6	Cl 300 3.2	Ar 363
K 100 0.8	Ca 141 1.0	Ga 138 1.6	Ge 182 1.9	As 226 2.0	Se 225 2.5	Br 273 2.9	Kr 323
Rb 96 0.8	Sr 131 1.0	In 133 1.7	Sn 169 1.8	Sb 199 2.1	Te 208 2.3	I 241 2.7	Xe 280
Cs 90 0.7	Ba 120 0.9	Tl 141 1.8	Pb 171 1.8	Bi 168 1.9	Po 194 2.0	At 2.2	Rn 248
Fr 0.7	Ra 122 0.9	*Arbitrary scale based on fluorine = 4.0					

Reference Table L

Relative Strength of Acids in Aqueous Solution at 1 atm and 298 K

Conjugate Pairs		K_a
Acid	*Base*	
$HI = H^+ + I^-$		very large
$HBr = H^+ + Br^-$		very large
$HCl = H^+ + Cl^-$		very large
$HNO_3 = H^+ + NO_3^-$		very large
$H_2SO_4 = H^+ + HSO_4^-$		large
$H_2O + SO_2 = H^+ + HSO_3^-$		1.5×10^{-2}
$HSO_4^- = H^+ + SO_4^{2-}$		1.2×10^{-2}
$H_3PO_4 = H^+ + H_2PO_4^-$		7.5×10^{-3}
$Fe(H_2O)_6^{3+} = H^+ + Fe(H_2O)_5(OH)^{2+}$		8.9×10^{-4}
$HNO_2 = H^+ + NO_2^-$		4.6×10^{-4}
$HF = H^+ + F^-$		3.5×10^{-4}
$Cr(H_2O)_6^{3+} = H^+ + Cr(H_2O)_5(OH)^{2+}$		1.0×10^{-4}
$CH_3COOH = H^+ + CH_3COO^-$		1.8×10^{-5}
$Al(H_2O)_6^{3+} = H^+ + Al(H_2O)_5(OH)^{2+}$		1.1×10^{-5}
$H_2O + CO_2 = H^+ + HCO_3^-$		4.3×10^{-7}
$HSO_3^- = H^+ + SO_3^{2-}$		1.1×10^{-7}
$H_2S = H^+ + HS^-$		9.5×10^{-8}
$H_2PO_4^- = H^+ + HPO_4^{2-}$		6.2×10^{-8}
$NH_4^+ = H^+ + NH_3$		5.7×10^{-10}
$HCO_3^- = H^+ + CO_3^{2-}$		5.6×10^{-11}
$HPO_4^{2-} = H^+ + PO_4^{3-}$		2.2×10^{-13}
$HS^- = H^+ + S^{2-}$		1.3×10^{-14}
$H_2O = H^+ + OH^-$		1.0×10^{-14}
$OH^- = H^+ + O^{2-}$		$< 10^{-36}$
$NH_3 = H^+ + NH_2^-$		very small

Note: $H^+(aq) = H_3O^+$

Sample equation: $HI + H_2O = H_3O^+ + I^-$

Reference Table M

Constants for Various Equilibria at 1 atm and 298 K

$H_2O(\ell) = H^+(aq) + OH^-(aq)$	$K_w = 1.0 \times 10^{-14}$
$H_2O(\ell) + H_2O(\ell) = H_3O^+(aq) + OH^-(aq)$	$K_w = 1.0 \times 10^{-14}$
$CH_3COO^-(aq) + H_2O(\ell) = CH_3COOH(aq) + OH^-(aq)$	$K_b = 5.6 \times 10^{-10}$
$Na^+F^-(aq) + H_2O(\ell) = Na^+(OH)^- + HF(aq)$	$K_b = 1.5 \times 10^{-1.1}$
$NH_3(aq) + H_2O(\ell) = NH_4^+(aq) + OH^-(aq)$	$K_b = 1.8 \times 10^{-5}$
$CO_3^{2-}(aq) + H_2O(\ell) = HCO_3^-(aq) + OH^-(aq)$	$K_b = 1.8 \times 10^{-4}$
$Ag(NH_3)_2^+(aq) = Ag^+(aq) + 2NH_3(aq)$	$K_{eq} = 8.9 \times 10^{-8}$
$N_2(g) + 3H_2(g) = 2NH_3(g)$	$K_{eq} = 6.7 \times 10^5$
$H_2(g) + I_2(g) = 2HI(g)$	$K_{eq} = 3.5 \times 10^{-1}$

Compound	K_{sp}	Compound	K_{sp}
AgBr	5.0×10^{-13}	Li_2CO_3	2.5×10^{-2}
AgCl	1.8×10^{-10}	$PbCl_2$	1.6×10^{-5}
Ag_2CrO_4	1.1×10^{-12}	$PbCO_3$	7.4×10^{-14}
AgI	8.3×10^{-17}	$PbCrO_4$	2.8×10^{-13}
$BaSO_4$	1.1×10^{-10}	PbI_2	7.1×10^{-9}
$CaSO_4$	9.1×10^{-6}	$ZnCO_3$	1.4×10^{-11}

Reference Table N

Standard Electrode Potentials

Ionic Concentrations 1 M Water at 298 K, 1 atm	
Half-Reaction	E^0 (volts)
$F_2(g) + 2e^- \rightarrow 2F^-$	+2.87
$8H^+ + MnO_4^- + 5e^- \rightarrow Mn^{2+} + 4H_2O$	+1.51
$Au^{3+} + 3e^- \rightarrow Au(s)$	+1.50
$Cl_2(g) + 2e^- \rightarrow 2Cl^-$	+1.36
$14H^+ + Cr_2O_7^{2-} + 6e^- \rightarrow 2Cr^{3+} + 7H_2O$	+1.23
$4H^+ + O_2(g) + 4e^- \rightarrow 2H_2O$	+1.23
$4H^+ + MnO_2(s) + 2e^- \rightarrow Mn^{2+} + 2H_2O$	+1.22
$Br_2(l) + 2e^- \rightarrow 2Br^-$	+1.09
$Hg^{2+} + 2e^- \rightarrow Hg(l)$	+0.85
$Ag^+ + e^- \rightarrow Ag(s)$	+0.80
$Hg_2^{2+} + 2e^- \rightarrow 2Hg(l)$	+0.80
$Fe^{3+} + e^- \rightarrow Fe^{2+}$	+0.77
$I_2(s) + 2e^- \rightarrow 2I^-$	+0.54
$Cu^+ + e^- \rightarrow Cu(s)$	+0.52
$Cu^{2+} + 2e^- \rightarrow Cu(s)$	+0.34
$4H^+ + SO_4^{2-} + 2e^- \rightarrow SO_2(aq) + 2H_2O$	+0.17
$Sn^{4+} + 2e^- \rightarrow Sn^{2+}$	+0.15
$2H^+ + 2e^- \rightarrow H_2(g)$	0.00
$Pb^{2+} + 2e^- \rightarrow Pb(s)$	−0.13
$Sn^{2+} + 2e^- \rightarrow Sn(s)$	−0.14
$Ni^{2+} + 2e^- \rightarrow Ni(s)$	−0.26
$Co^{2+} + 2e^- \rightarrow Co(s)$	−0.28
$Fe^{2+} + 2e^- \rightarrow Fe(s)$	−0.45
$Cr^{3+} + 3e^- \rightarrow Cr(s)$	−0.74
$Zn^{2+} + 2e^- \rightarrow Zn(s)$	−0.76
$2H_2O + 2e^- \rightarrow 2OH^- + H_2(g)$	−0.83
$Mn^{2+} + 2e^- \rightarrow Mn(s)$	−1.19
$Al^{3+} + 3e^- \rightarrow Al(s)$	−1.66
$Mg^{2+} + 2e^- \rightarrow Mg(s)$	−2.37
$Na^+ + e^- \rightarrow Na(s)$	−2.71
$Ca^{2+} + 2e^- \rightarrow Ca(s)$	−2.87
$Sr^{2+} + 2e^- \rightarrow Sr(s)$	−2.89
$Ba^{2+} + 2e^- \rightarrow Ba(s)$	−2.91
$Cs^+ + e^- \rightarrow Cs(s)$	−2.92
$K^+ + e^- \rightarrow K(s)$	−2.93
$Rb^+ + e^- \rightarrow Rb(s)$	−2.98
$Li^+ + e^- \rightarrow Li(s)$	−3.04

Reference Table O

Vapor Pressure of Water

°C	torr (mm Hg)	°C	torr (mm Hg)
0	4.6	26	25.2
5	6.5	27	26.7
10	9.2	28	28.3
15	12.8	29	30.0
16	13.6	30	31.8
17	14.5	40	55.3
18	15.5	50	92.5
19	16.5	60	149.4
20	17.5	70	233.7
21	18.7	80	355.1
22	19.8	90	525.8
23	21.1	100	760.0
24	22.4	105	906.1
25	23.8	110	1074.6

Reference Table P

Radii of Atoms

KEY

	Symbol	F
Covalent Radius, Å →		0.64
Atomic Radius in Metals, Å →		(-)
Van der Waals Radius, Å →		1.35

A dash (–) indicates data are not available.

Element	Covalent Radius, Å	Atomic Radius in Metals, Å	Van der Waals Radius, Å
H	0.37	(-)	1.2
He	(-)	(-)	1.22
Li	1.23	1.52	(-)
Be	0.89	1.13	(-)
B	0.88	(-)	2.08
C	0.77	(-)	1.85
N	0.70	(-)	1.54
O	0.66	(-)	1.40
F	0.64	(-)	1.35
Ne	(-)	(-)	1.60
Na	1.57	1.54	2.31
Mg	1.36	1.60	(-)
Al	1.25	1.43	(-)
Si	1.17	(-)	2.0
P	1.10	(-)	1.90
S	1.04	(-)	1.85
Cl	0.99	(-)	1.81
Ar	(-)	(-)	1.91
K	2.03	2.27	2.31
Ca	1.74	1.97	(-)
Sc	1.44	1.61	(-)
Ti	1.32	1.45	(-)
V	1.22	1.32	(-)
Cr	1.17	1.25	(-)
Mn	1.17	1.24	(-)
Fe	1.17	1.24	(-)
Co	1.16	1.25	(-)
Ni	1.15	1.25	(-)
Cu	1.17	1.28	(-)
Zn	1.25	1.33	(-)
Ga	1.25	1.22	(-)
Ge	1.22	1.23	(-)
As	1.21	(-)	2.0
Se	1.17	(-)	2.0
Br	1.14	(-)	1.95
Kr	1.89	(-)	1.98
Rb	2.16	2.48	2.44
Sr	1.92	2.15	(-)
Y	1.62	1.81	(-)
Zr	1.45	1.60	(-)
Nb	1.34	1.43	(-)
Mo	1.29	1.36	(-)
Tc	1.27	1.36	(-)
Ru	1.24	1.33	(-)
Rh	1.25	1.35	(-)
Pd	1.28	1.38	(-)
Ag	1.34	1.44	(-)
Cd	1.41	1.49	(-)
In	1.50	1.63	(-)
Sn	1.40	1.41	(-)
Sb	1.41	(-)	2.2
Te	1.37	(-)	2.20
I	1.33	(-)	2.15
Xe	2.09	(-)	(-)
Cs	2.35	2.65	2.62
Ba	1.98	2.17	(-)
La–Lu			
Hf	1.44	1.56	(-)
Ta	1.34	1.43	(-)
W	1.30	1.37	(-)
Re	1.28	1.37	(-)
Os	1.26	1.34	(-)
Ir	1.26	1.36	(-)
Pt	1.29	1.38	(-)
Au	1.34	1.44	(-)
Hg	1.44	1.60	(-)
Tl	1.55	1.70	(-)
Pb	1.54	1.75	(-)
Bi	1.52	1.55	(-)
Po	1.53	1.67	(-)
At	(-)	(-)	(-)
Rn	2.14	(-)	(-)
Fr	(-)	2.7	(-)
Ra	(-)	2.20	(-)
Ac–Lr			

Lanthanide series (La–Lu)

Element	Covalent Radius, Å	Atomic Radius in Metals, Å	Van der Waals Radius, Å
La	1.69	1.88	(-)
Ce	1.65	1.83	(-)
Pr	1.65	1.83	(-)
Nd	1.64	1.82	(-)
Pm	(-)	1.81	(-)
Sm	1.66	1.80	(-)
Eu	1.85	2.04	(-)
Gd	1.61	1.80	(-)
Tb	1.59	1.78	(-)
Dy	1.59	1.77	(-)
Ho	1.58	1.77	(-)
Er	1.57	1.76	(-)
Tm	1.56	1.75	(-)
Yb	1.70	1.94	(-)
Lu	1.56	1.73	(-)

Actinide series (Ac–Lr)

Element	Covalent Radius, Å	Atomic Radius in Metals, Å	Van der Waals Radius, Å
Ac	(-)	1.88	(-)
Th	(-)	1.80	(-)
Pa	(-)	1.61	(-)
U	(-)	1.39	(-)
Np	(-)	1.31	(-)
Pu	(-)	1.51	(-)
Am	(-)	1.84	(-)
Cm	(-)	(-)	(-)
Bk	(-)	(-)	(-)
Cf	(-)	(-)	(-)
Es	(-)	(-)	(-)
Fm	(-)	(-)	(-)
Md	(-)	(-)	(-)
No	(-)	(-)	(-)
Lr	(-)	(-)	(-)

Periodic Table of the Elements

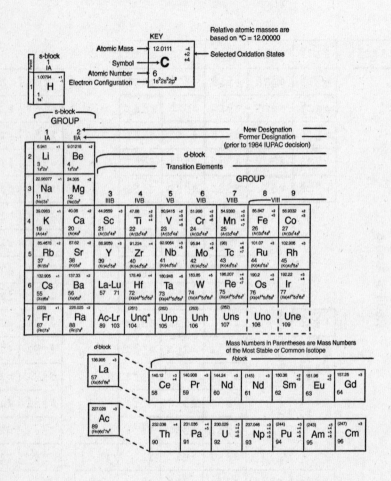

p-block
Group

			13 IIIA	14 IVA	15 VA	16 VIA	17 VIIA	18 0
			10.81 +3 **B** 5 $1s^22s^22p^1$	12.0111 -4,+2,+4 **C** 6 $1s^22s^22p^2$	14.0067 -3 **N** 7 $1s^22s^22p^3$	15.9994 -2 **O** 8 $1s^22s^22p^4$	18.998403 -1 **F** 9 $1s^22s^22p^5$	20.179 0 **Ne** 10 $1s^22s^22p^6$
10	11 IB	12 IIB	26.98154 +3 **Al** 13 $(Ne)3s^23p^1$	28.0655 -4,+2,+4 **Si** 14 $(Ne)3s^23p^2$	30.97376 -3,+3,+5 **P** 15 $(Ne)3s^23p^3$	32.06 -2,+4,+6 **S** 16 $(Ne)3s^23p^4$	35.453 -1,+1,+3,+5,+7 **Cl** 17 $(Ne)3s^23p^5$	39.948 0 **Ar** 18 $(Ne)3s^23p^6$
58.69 +2,+3 **Ni** 28 $(Ar)3d^84s^2$	63.546 +1,+2 **Cu** 29 $(Ar)3d^{10}4s^1$	65.39 +2 **Zn** 30 $(Ar)3d^{10}4s^2$	69.72 +3 **Ga** 31 $(Ar)3d^{10}4s^24p^1$	72.59 +2,+4 **Ge** 32 $(Ar)3d^{10}4s^24p^2$	74.9216 -3,+3,+5 **As** 33 $(Ar)3d^{10}4s^24p^3$	78.96 -2,+4,+6 **Se** 34 $(Ar)3d^{10}4s^24p^4$	79.904 -1,+1,+5 **Br** 35 $(Ar)3d^{10}4s^24p^5$	83.80 0 **Kr** 36 $(Ar)3d^{10}4s^24p^6$
106.42 +2 **Pd** 46 $(Kr)4d^{10}5s^0$	107.868 +1 **Ag** 47 $(Kr)4d^{10}5s^1$	112.41 +2 **Cd** 48 $(Kr)4d^{10}5s^2$	114.82 +3 **In** 49 $(Kr)4d^{10}5s^25p^1$	118.71 +2,+4 **Sn** 50 $(Kr)4d^{10}5s^25p^2$	121.75 +3,+5 **Sb** 51 $(Kr)4d^{10}5s^25p^3$	127.60 -2,+4,+6 **Te** 52 $(Kr)4d^{10}5s^25p^4$	126.905 -1,+1,+5,+7 **I** 53 $(Kr)4d^{10}5s^25p^5$	131.29 0,+2,+4,+6 **Xe** 54 $(Kr)4d^{10}5s^25p^6$
195.08 +2,+4 **Pt** 78 $(Xe)4f^{14}5d^96s^1$	196.967 +1,+3 **Au** 79 $(Xe)4f^{14}5d^{10}6s^1$	200.59 +1,+2 **Hg** 80 $(Xe)4f^{14}5d^{10}6s^2$	204.383 +1,+3 **Tl** 81 $(Xe)4f^{14}5d^{10}6s^26p^1$	207.2 +2,+4 **Pb** 82 $(Xe)4f^{14}5d^{10}6s^26p^2$	208.980 +3 **Bi** 83 $(Xe)4f^{14}5d^{10}6s^26p^3$	(209) +2,+4 **Po** 84 $(Xe)4f^{14}5d^{10}6s^26p^4$	(210) **At** 85 $(Xe)4f^{14}5d^{10}6s^26p^5$	(222) 0 **Rn** 86 $(Xe)4f^{14}5d^{10}6s^26p^6$

* The systematic names and symbols for elements of atomic numbers greater than 103 will be used until the approval of trivial names by IUPAC.

f-block

158.925 +3 **Tb** 65	162.50 +3 **Dy** 66	164.930 +3 **Ho** 67	167.26 +3 **Er** 68	168.934 +3 **Tm** 69	173.04 +3 **Yb** 70	174.967 +3 **Lu** 71	Lanthanoid Series

(247) +3 **Bk** 97	(251) +3 **Cf** 98	(252) **Es** 99	(257) **Fm** 100	(258) **Md** 101	(259) **No** 102	(260) **Lr** 103	Actinoid Series

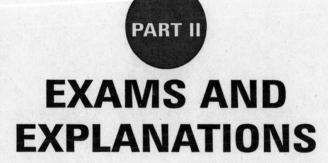

PART II

EXAMS AND EXPLANATIONS

EXAMINATION
JUNE 1992

PART 1: *Answer all 56 questions in this part.* [65]

DIRECTIONS **(1–56):** *For each statement or question, select the word or expression that, of those given, best completes the statement or answers the question. Record your answer on the separate answer sheet provided.*

1 Which substance can be decomposed by a chemical change?
1 beryllium   3 methanol
2 boron   4 magnesium

2 The particles in a crystalline solid are arranged
1 randomly and far apart
2 randomly and close together
3 regularly and far apart
4 regularly and close together

3 The strongest intermolecular forces of attraction exist in a liquid whose heat of vaporization is
(1) 100 cal/g   (3) 300 cal/g
(2) 200 cal/g   (4) 400 cal/g

4 If the pressure on the surface of water in the liquid state is 355.1 torr, the water will boil at
(1) 0°C   (3) 80°C
(2) 40°C   (4) 100°C

5 When a substance was dissolved in water, the temperature of the water increased. This process is described as
   1 endothermic, with the release of energy
   2 endothermic, with the absorption of energy
   3 exothermic, with the release of energy
   4 exothermic, with the absorption of energy

6 Compared to the entire atom, the nucleus of the atom is
   1 smaller and contains most of the atom's mass
   2 smaller and contains little of the atom's mass
   3 larger and contains most of the atom's mass
   4 larger and contains little of the atom's mass

7 All isotopes of a given element must have the same
   1 atomic mass
   2 atomic number
   3 mass number
   4 number of neutrons

8 In the equation $^{234}_{90}\text{Th} \rightarrow {}^{234}_{91}\text{Pa} + X$, which particle is represented by X?
   (1) $^{0}_{-1}\text{e}$ (3) $^{1}_{1}\text{H}$

   (2) $^{4}_{2}\text{He}$ (4) $^{0}_{+1}\text{e}$

9 Four valence electrons of an atom in the ground state would occupy the
   (1) s sublevel, only
   (2) p sublevel, only
   (3) s and p sublevels, only
   (4) s, p, and d sublevels

10 Which particle has a mass of approximately one atomic mass unit and a unit positive charge?
   1 a neutron   3 a beta particle
   2 a proton   4 an alpha particle

11 An atom of carbon-14 contains
   (1) 8 protons, 6 neutrons, and 6 electrons
   (2) 6 protons, 6 neutrons, and 8 electrons
   (3) 6 protons, 8 neutrons, and 8 electrons
   (4) 6 protons, 8 neutrons, and 6 electrons

12 In an atom that has an electron configuration of $1s^2 2s^2 2p^3$, what is the total number of electrons in its sublevel of highest energy?
   (1) 1   (3) 3
   (2) 2   (4) 4

13 The correct electron dot formula for hydrogen chloride is
   (1) H: Cl   (3) :$\ddot{\text{H}}$: Cl

   (2) :$\ddot{\text{H}}$: Cl   (4) :H: $\ddot{\text{Cl}}$:

14 What is the total mass of oxygen in 1.00 mole of $Al_2(CrO_4)_3$?
   (1) 192 g   (3) 64.0 g
   (2) 112 g   (4) 48.0 g

15 Which molecule is nonpolar and has a symmetrical shape?
(1) HCl
(2) CH_4
(3) H_2O
(4) NH_3

16 What is the correct formula for ammonium carbonate?
(1) $NH_4(CO_3)_2$
(2) NH_4CO_3
(3) $(NH_4)_2(CO_3)_2$
(4) $(NH_4)_2CO_3$

17 What type of bonding is present within a network solid?
1 hydrogen
2 covalent
3 ionic
4 metallic

18 The table below lists four different chemical bonds and the amount of energy released when 1 mole of each of the bonds is formed.

Bond	Energy Released in Formation (kcal/mole)
H—F	135
H—Cl	103
H—Br	87
H—I	71

Which bond is the most stable?
(1) H—F
(2) H—Cl
(3) H—Br
(4) H—I

19 Which element is in Group 2 (IIA) and Period 7 of the Periodic Table?
    1 magnesium   3 radium
    2 manganese   4 radon

20 Which ion has the largest radius?
    (1) I$^-$ (3) Br$^-$
    (2) Cl$^-$ (4) F$^-$

21 An atom in the ground state contains 8 valence electrons. This atom is classified as a
    1 metal   3 noble gas
    2 semimetal   4 halogen

22 Nonmetals in the solid state are poor conductors of heat and tend to
    1 be brittle
    2 be malleable
    3 have a shiny luster
    4 have good electrical conductivity

23 Which group contains elements composed of diatomic molecules at STP?
    (1) 11 (IB)   (3) 7 (VIIB)
    (2) 2 (IIA)   (4) 17 (VIIA)

24 Which metal atoms can form ionic bonds by losing electrons from both the outermost and next to outermost principal energy levels?
    (1) Fe   (3) Mg
    (2) Pb   (4) Ca

25 Which species contains the greatest percent by mass of hydrogen?

(1) OH^- (3) H_3O^+

(2) H_2O (4) H_2O_2

26 Given the reaction:

$$CH_4(g) + 2O_2(g) \rightarrow CO_2(g) + 2H_2O(g)$$

How many moles of oxygen are needed for the complete combustion of 3.0 moles of $CH_4(g)$?

(1) 6.0 moles   (3) 3.0 moles

(2) 2.0 moles   (4) 4.0 moles

27 How many moles of $N_2(g)$ molecules would contain exactly 4.0 moles of nitrogen atoms?

(1) 1.0   (3) 3.0

(2) 2.0   (4) 4.0

28 What mass contains 6.0×10^{23} atoms?

(1) 6.0 g of carbon   (3) 3.0 g of helium

(2) 16 g of sulfur   (4) 28 g of silicon

29 Which is a homogeneous mixture?

(1) $I_2(s)$ (3) $HCl(g)$

(2) $I_2(l)$ (4) $HCl(aq)$

30 When a catalyst is added to a chemical reaction, there is a change in the

1 heat of reaction

2 rate of reaction

3 potential energy of the reactants

4 potential energy of the products

31 A piece of Mg(s) ribbon is held in a bunsen burner flame and begins to burn according to the equation $2Mg(s) + O_2(g) \rightarrow 2MgO(s)$. The reaction begins because the reactants
1 are activated by heat from the bunsen burner flame
2 are activated by heat from the burning magnesium
3 underwent an increase in entropy
4 underwent a decrease in entropy

32 Raising the temperature speeds up the rate of a chemical reaction by increasing
1 the effectiveness of the collisions, only
2 the frequency of the collisions, only
3 both the effectiveness and the frequency of the collisions
4 neither the effectiveness nor the frequency of the collisions

33 Which tendencies favor a spontaneous reaction?
1 decreasing enthalpy and decreasing entropy
2 decreasing enthalpy and increasing entropy
3 increasing enthalpy and decreasing entropy
4 increasing enthalpy and increasing entropy

34 Which formula represents a salt?
(1) KOH
(2) KCl
(3) CH_3OH
(4) CH_3COOH

35 What is the H_3O^+ ion concentration of a solution whose OH^- ion concentration is 1×10^{-3} M?
(1) 1×10^{-4} M   (3) 1×10^{-11} M
(2) 1×10^{-7} M   (4) 1×10^{-14} M

36 Which of the following acids is the weakest?
(1) H_2S (3) H_3PO_4
(2) HF   (4) HNO_2

37 Given the reaction at equilibrium:

$$NH_4 + OH^- \rightleftarrows H_2O + NH_3$$

Which species is the Brönsted-Lowry acid in the forward reaction?
(1) NH_3 (3) OH^-
(2) H_2O (4) NH_4^+

38 To neutralize 1 mole of sulfuric acid, 2 moles of sodium hydroxide are required. How many liters of 1 M NaOH are needed to exactly neutralize 1 liter of 1 M H_2SO_4?
(1) 1   (3) 0.5
(2) 2   (4) 4

39 What is the conjugate base of NH_3?
(1) NH_4^+ (3) NO_3^-
(2) NH_2^- (4) NO_2^-

40 An aqueous solution of an ionic compound turns red litmus blue, conducts electricity, and reacts with an acid to form a salt and water. This compound could be
(1) HCl   (3) KNO_3
(2) NaI   (4) LiOH

41 Oxygen will have a positive oxidation number when combined with
1 fluorine   3 bromine
2 chlorine   4 iodine

42 In the reaction

$$2Al(s) + 3Cu^{2+}(aq) \rightarrow 2Al^{3+}(aq) + 3Cu(s),$$

the Al(s)
1 gains protons
2 loses protons
3 gains electrons
4 loses electrons

43 The purpose of a salt bridge in an electrochemical cell is to
1 allow for the flow of molecules between the solutions
2 allow for the flow of ions between the solutions
3 prevent the flow of molecules between the solutions
4 prevent the flow of ions between the solutions

44 In the reaction Pb + 2Ag$^+$ → Pb^{2+} + 2 Ag, the Ag$^+$ is
   1 reduced, and the oxidation number changes
     from +1 to 0
   2 reduced, and the oxidation number changes
     from +2 to 0
   3 oxidized, and the oxidation number changes
     from 0 to +1
   4 oxidized, and the oxidation number changes
     from +1 to 0

45 Which half-reaction correctly represents oxidation?
   (1) Mg + 2e$^-$ → Mg^{2+}
   (2) Mg^{2+} + 2e$^-$ → Mg
   (3) Mg^{2+} → Mg + 2e$^-$
   (4) Mg → Mg^{2+} + 2e$^-$

46 When the equation
   __Pb^{2+} + __Au^{3+} → __Pb^{4+} + __Au is correctly bal-
   anced using the smallest whole-number coefficients, the
   coefficient of the Pb^{2+} will be
   (1) 1   (3) 3
   (2) 2   (4) 4

47 Which alcohol reacts with C$_2$H$_5$COOH to produce the
   ester C$_2$H$_5$COOC$_2$H$_5$?
   (1) CH$_3$OH   (3) C$_3$H$_7$OH
   (2) C$_2$H$_5$OH   (4) C$_4$H$_9$OH

48 A compound with the molecular formula C_7H_8 could be a member of the hydrocarbon series which has the general formula

(1) C_nH_{2n} (3) C_nH_{2n-6}

(2) C_nH_{2n-2} (4) C_nH_{2n+2}

49 Given the equation:

$$C_6H_{12}O_6 \xrightarrow{\text{zymase}} 2C_2H_5OH + 2CO_2$$

The reaction represented by this equation is called

1 esterification   3 fermentation

2 saponification   4 polymerization

50 Which structural formula represents an isomer of

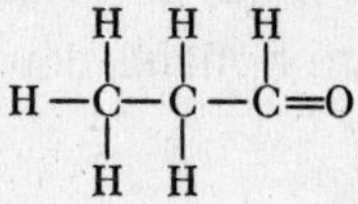

(1) 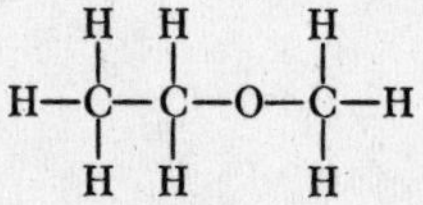   (3)

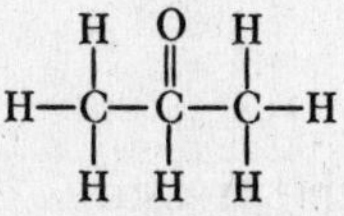

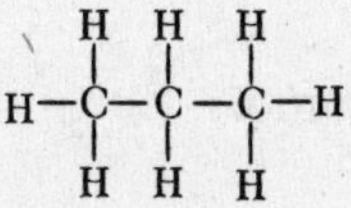

(2)   (4)

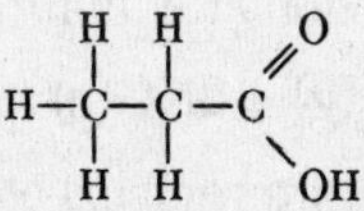

51 Which compound can undergo an addition reaction?

(1) CH_4 (3) C_3H_8

(2) C_2H_4 (4) C_4H_{10}

Note that questions 52 through 56 have only three choices.

52 A closed system is shown in the diagram below.

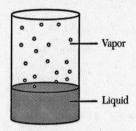

The rate of vapor formation at equilibrium is
1 less than the rate of liquid formation
2 greater than the rate of liquid formation
3 equal to the rate of liquid formation

53 If the pressure on a given mass of gas in a closed system is increased and the temperature remains constant, the volume of the gas will
1 decrease
2 increase
3 remain the same

54 Given the reaction at equilibrium:

$$N_2(g) + O_2(g) = 2NO(g)$$

If the temperature remains constant and the pressure increases, the number of moles of NO(g) will
1 decrease
2 increase
3 remain the same

55 As a chemical bond forms between two hydrogen atoms, the potential energy of the atoms
1 decreases
2 increases
3 remains the same

56 As the elements in Period 3 of the Periodic Table are considered in order of increasing atomic number, the ability of each successive element to act as a reducing agent
1 decreases
2 increases
3 remains the same

PART 2: *This part consists of twelve groups. Choose seven of these twelve groups. Be sure to answer all questions in each group chosen. Write the answers to these questions on the separate answer sheet provided.* [35]

GROUP 1—Matter and Energy

If you choose this group, be sure to answer questions **57–61.**

57 Compared to the average kinetic energy of 1 mole of water at 0°C, the average kinetic energy of 1 mole of water at 298 K is
1 the same, and the number of molecules is the same
2 the same, but the number of molecules is greater
3 greater, and the number of molecules is greater
4 greater, but the number of molecules is the same

58 Which formula represents a binary compound?
   (1) Ne   (3) C_3H_8
   (2) Br_2 (4) H_2SO_4

59 At standard pressure, which element at 25°C could undergo a change of phase when the temperature is decreased?
   1 aluminum   3 silicon
   2 chlorine   4 sulfur

60 Which change results in a release of energy?
   1 the melting of $H_2O(s)$
   2 the boiling of $H_2O(\ell)$
   3 the evaporation of $H_2O(\ell)$
   4 the condensation of $H_2O(g)$

61 A cylinder is filled with 2.00 moles of nitrogen, 3.00 moles of argon, and 5.00 moles of helium. If the gas mixture is at STP, what is the partial pressure of the argon?
   (1) 152 torr   (3) 380. torr
   (2) 228 torr   (4) 650. torr

GROUP 2—Atomic Structure

If you choose this group, be sure to answer questions 62–66.

62 The total number of orbitals in a *d* sublevel is
   (1) 1   (3) 3
   (2) 5   (4) 7

63 Compared to ^{37}K, the isotope ^{42}K has a
   1 shorter half-life and the same decay mode
   2 shorter half-life and a different decay mode
   3 longer half-life and the same decay mode
   4 longer half-life and a different decay mode

64 A sample of ^{131}I decays to 1.0 gram in 40. days. What was the mass of the original sample?
(1) 8.0 g   (3) 32 g
(2) 16 g   (4) 4.0 g

65 What is the total number of occupied sublevels in an atom of chlorine in the ground state?
(1) 1   (3) 3
(2) 5   (4) 9

66 The characteristic bright-line spectrum of sodium is produced when its electrons
1 return to lower energy levels
2 jump to higher energy levels
3 are lost by the neutral atoms
4 are gained by the neutral atoms

GROUP 3—Bonding

If you choose this group, be sure to answer questions 67–71.

67 A substance that has a melting point of 1074 K conducts electricity when dissolved in water, but does *not* conduct electricity in the solid phase. The substance is most likely
1 an ionic solid   3 a metallic solid
2 a network solid   4 a molecular solid

68 Which quantity is represented by the symbol Ne?
(1) 1 gram of neon
(2) 1 liter of neon
(3) 1 mole of neon
(4) 1 atomic mass unit of neon

69 The diagram below represents a water molecule

This molecule is best described as
1 polar with polar covalent bonds
2 polar with nonpolar covalent bonds
3 nonpolar with polar covalent bonds
4 nonpolar with nonpolar covalent bonds

70 The diagrams below represent an ionic crystal being dissolved in water.

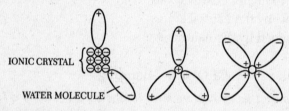

IONIC CRYSTAL

WATER MOLECULE

According to the diagrams, the dissolving process takes place by
1 hydrogen bond formation
2 network bond formation
3 van der Waals attractions
4 molecule-ion attractions

71 When the equation
$$\_Al_2(SO_4)_3 + \_ZnCl_2 \rightarrow \_AlCl_3 + \_ZnSO_4$$
is correctly balanced using the smallest whole-number coefficients, the sum of the coefficients is
(1) 9   (3) 5
(2) 8   (4) 4

GROUP 4—Periodic Table

If you choose this group, be sure to answer questions 72–76.

72 Which aqueous salt solution has a color?
(1) $BaSO_4(aq)$ (3) $SrSO_4(aq)$
(2) $CuSO_4(aq)$ (4) $MgSO_4(aq)$

73 Which category is composed of elements that have both positive and negative oxidation states?
1 the alkali metals
2 the transition metals
3 the halogens
4 the alkaline earths

74 As the atoms of the elements from atomic number 3 to atomic number 9 are considered in sequence from left to right on the Periodic Table, the covalent atomic radius of each successive atom is
1 smaller, and the nuclear charge is less
2 smaller, and the nuclear charge is greater
3 larger, and the nuclear charge is less
4 larger, and the nuclear charge is greater

75 Which element is so active chemically that it occurs naturally only in compounds?
1 potassium   3 copper
2 silver   4 sulfur

76 Which element is a solid at room temperature and standard pressure?
1 bromine   3 mercury
2 iodine   4 neon

GROUP 5—Mathematics of Chemistry

If you choose this group, be sure to answer questions **77–81.**

77 The temperature of 100 grams of water changes from 16°C to 20°C. What is the total number of calories of heat energy absorbed by the water?
(1) 25   (3) 100
(2) 40   (4) 400

78 When ethylene glycol (an antifreeze) is added to water, the boiling point of the water
1 decreases, and the freezing point decreases
2 decreases, and the freezing point increases
3 increases, and the freezing point decreases
4 increases, and the freezing point increases

79 A sample of a compound contains 24 grams of carbon and 64 grams of oxygen. What is the empirical formula of this compound?
(1) CO   (3) C_2O_2
(2) CO_2 (4) C_2O_4

80 Which gas has approximately the same density as C_2H_6 at STP?
(1) NO   (3) H_2S
(2) NH_3 (4) SO_2

81 At a temperature of 273 K, a 400.-milliliter gas sample has a pressure of 760. millimeters of mercury. If the pressure is changed to 380. millimeters of mercury, at which temperature will this gas sample have a volume of 551 milliliters?
(1) 100 K   (3) 273 K
(2) 188 K   (4) 546 K

GROUP 6—Kinetics and Equilibrium

If you choose this group, be sure to answer questions 82–86.

82 According to Reference Table E, which salt would have the smallest K_{sp} value?
(1) $AlBr_3$ (3) $NaBr$
(2) $PbBr_2$ (4) $AgBr$

83 Given the solution at equilibrium:

$$CaF_2(s) \rightleftarrows Ca^{2+}(aq) + 2F^-(aq)$$

What is the solubility product expression (K_{sp})?
(1) $K_{sp} = [Ca^{2+}][F^-]$
(2) $K_{sp} = [Ca^{2+}][2F^-]$
(3) $K_{sp} = [Ca^{2+}][F^-]^2$
(4) $K_{sp} = [Ca^{2+}]^2[F^-]$

84 Based on Reference Table G, which compound will form spontaneously from its elements?
1 carbon dioxide (g)
2 nitrogen (II) oxide (g)
3 ethene (g)
4 ethyne (g)

85 Given the reaction:

$$2Na(s) + 2H_2O(\ell) \rightarrow$$
$$2Na^+(aq) + 2OH^-(aq) + H_2(g)$$

This reaction goes to completion because one of the products formed is
1 an insoluble base   3 a precipitate
2 a soluble base   4 a gas

86 Given the reaction at equilibrium:

$$PbCl_2(s) \rightleftarrows Pb^{2+}(aq) + 2Cl^-(aq)$$

When KCl(s) is added to the system, the equilibrium shifts to the
1 right, and the concentration of $Pb^{2+}(aq)$ ions decreases
2 right, and the concentration of $Pb^{2+}(aq)$ ions increases
3 left, and the concentration of $Pb^{2+}(aq)$ ions decreases
4 left, and the concentration of $Pb^{2+}(aq)$ ions increases

GROUP 7—Acids and Bases

If you choose this group, be sure to answer questions 87–91.

87 Given the reaction:

$$HF(aq) \rightleftarrows H^+(aq) + F^-(aq)$$

Which expression represents the equilibrium constant (K_a) for the acid HF?

(1) $K_a = \dfrac{[H^+][F^-]}{[HF]}$ (3) $K_a = \dfrac{[HF]}{[H^+][F^-]}$

(2) $K_a = [H^+][F^-]$ (4) $K_a = 2[HF]$

88 According to Reference Table *L*, which substance is amphoteric (amphiprotic)?
(1) HNO_3 (3) HCO_3^-
(2) NO_2^- (4) HF

89 Which substance can act as an Arrhenius acid in aqueous solution?
(1) NaI   (3) LiH
(2) HI   (4) NH_3

90 Which of the following metals will react most readily
with HCl(aq) to release hydrogen gas?
1 aluminum   3 silver
2 copper   4 gold

91 The diagram below illustrates an apparatus used to test
the conductivity of various solutions.

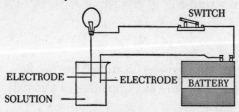

When the switch is closed, which of the following
1-molar solutions would cause the bulb to glow most
brightly?
1 ammonia   3 carbonic acid
2 acetic acid   4 sulfuric acid

GROUP 8—Redox and Electrochemistry
If you choose this group, be sure to answer questions 92–96.

92 Given the reaction:

$$Zn(s) + Br_2(aq) \rightarrow Zn^{2+}(aq) + 2Br^-(aq)$$

What is the net cell potential (E^0) for the overall reaction?
(1) +0.76 V   (3) +1.85 V
(2) −1.09 V   (4) 0.00 V

93 Which half-cell reaction serves as the arbitrary standard used to determine the standard electrode potentials?
(1) $Na^+ + e^- \rightarrow Na(s)$
(2) $Ag^+ + e^- \rightarrow Ag(s)$
(3) $F_2(g) + 2e^- \rightarrow 2F^-$
(4) $2H^+ + 2e^- \rightarrow H_2(g)$

94 According to Reference Table N, which element will react spontaneously with Al^{3+} at 298 K?
(1) Cu   (3) Li
(2) Au   (4) Ni

Base your answers to questions 95 and 96 on the diagram below, which represents the electroplating of a metal fork with Ag(s).

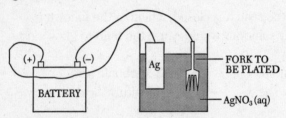

95 Which part of the electroplating system is provided by the fork?
1 the anode, which is the negative electrode
2 the cathode, which is the negative electrode
3 the anode, which is the positive electrode
4 the cathode, which is the positive electrode

96 Which equation represents the half-reaction that takes place at the fork?
(1) $Ag^+ + NO_3^- \rightarrow AgNO_3$
(2) $AgNO_3 \rightarrow Ag^+ + NO_3^-$
(3) $Ag^+ + e^- \rightarrow Ag(s)$
(4) $Ag(s) \rightarrow Ag^+ + e^-$

GROUP 9—Organic Chemistry

If you choose this group, be sure to answer questions 97–101.

97 Which type of reaction is used in the production of nylon?

   1 substitution   3 esterification

   2 saponification   4 polymerization

98 Which of the following hydrocarbons has the highest normal boiling point?

   1 butene   3 pentene

   2 ethene   4 propene

99 Which is the structural formula for 2-propanol?

(1)
$$H-\underset{\underset{H}{|}}{\overset{\overset{H}{|}}{C}}-\underset{\underset{H}{|}}{\overset{\overset{H}{|}}{C}}-\underset{\underset{H}{|}}{\overset{\overset{H}{|}}{C}}-OH$$

(2)
$$H-\underset{\underset{H}{|}}{\overset{\overset{H}{|}}{C}}-\underset{\underset{OH}{|}}{\overset{\overset{H}{|}}{C}}-\underset{\underset{H}{|}}{\overset{\overset{H}{|}}{C}}-H$$

(3)
$$H-\underset{\underset{H}{|}}{\overset{\overset{H}{|}}{C}}-\underset{\underset{H}{|}}{\overset{\overset{H}{|}}{C}}-\underset{\underset{H}{|}}{\overset{\overset{H}{|}}{C}}-\underset{\underset{H}{|}}{\overset{\overset{H}{|}}{C}}-OH$$

(4)
$$H-\underset{\underset{H}{|}}{\overset{\overset{H}{|}}{C}}-\underset{\underset{H}{|}}{\overset{\overset{H}{|}}{C}}-\underset{\underset{OH}{|}}{\overset{\overset{H}{|}}{C}}-\underset{\underset{H}{|}}{\overset{\overset{H}{|}}{C}}-H$$

100 Which is the general formula for organic acids?

(1) $R-C\overset{\displaystyle O}{\underset{\displaystyle H}{\diagup}}$

(3) $\overset{\displaystyle R_1}{\underset{\displaystyle R_2}{>}}C=O$

(2) $R-C\overset{\displaystyle O}{\underset{\displaystyle OH}{\diagup}}$

(4) R_1-O-R_2

101 Which is the structural formula for diethyl ether?

(1)
$$
\begin{array}{cccccc}
& H & H & & H & H \\
& | & | & & | & | \\
H- & C- & C- & O- & C- & C-H \\
& | & | & & | & | \\
& H & H & & H & H
\end{array}
$$

(2)
$$
\begin{array}{cccccc}
& H & H & & H & H \\
& | & | & & | & | \\
H- & C- & C- & C- & C- & C-H \\
& | & | & \| & | & | \\
& H & H & O & H & H
\end{array}
$$

(3)
$$
\begin{array}{ccc}
& H & H \\
& | & | \\
H- & C-O- & C-H \\
& | & | \\
& H & H
\end{array}
$$

(4)
$$
\begin{array}{ccc}
& H & H \\
& | & | \\
H- & C- C- & C-H \\
& | & | & | \\
& H & O & H
\end{array}
$$

GROUP 10—**Applications of Chemical Principles**

*If you choose this group, be sure to answer questions **102–106**.*

102 What is produced when sulfur is burned during the first step of the contact process?

1 sulfuric acid   3 sulfur trioxide
2 sulfur dioxide   4 pyrosulfuric acid

103 A common gaseous fuel that is often found with petroleum is

1 carbon monoxide   3 methane
2 carbon dioxide   4 ethene

104 Which method is commonly used to separate the components of petroleum into simpler substances by using their differences in boiling points?

1 fractional crystallization
2 fractional distillation
3 esterification
4 saponification

105 Given the reaction in a lead storage battery:

$$Pb + PbO_2 + 2H_2SO_4 \rightarrow 2PbSO_4 + 2H_2O$$

When the battery is being discharged, which change in the oxidation state of lead occurs?

(1) Pb is oxidized to Pb^{2+}.
(2) Pb is oxidized to Pb^{4+}.
(3) Pb^{2+} is reduced to Pb.
(4) Pb^{4+} is reduced to Pb.

106 Which metal oxide is most easily reduced by carbon?

1 aluminum   3 magnesium
2 iron   4 sodium

GROUP 11—Nuclear Chemistry

If you choose this group, be sure to answer questions **107–111.**

107 Which isotopic ratio needs to be determined when the age of ancient wooden objects is investigated?
   1 uranium-235 to uranium-238
   2 hydrogen-2 to hydrogen-3
   3 nitrogen-16 to nitrogen-14
   4 carbon-14 to carbon-12

108 Given the reaction: $^9_4Be + ^1_1H \rightarrow ^4_2He + X$
   Which species is represented by X?
   (1) ^8_3Li (3) ^8_5B

   (2) ^6_3Li (4) $^{10}_5B$

109 A particle accelerator has no effect on the velocity of
   1 an alpha particle   3 a neutron
   2 a beta particle   4 a proton

110 In a fusion reaction, a major problem related to causing the nuclei to fuse into a single nucleus is the
   1 small mass of the nuclei
   2 large mass of the nuclei
   3 attractions of the nuclei
   4 repulsions of the nuclei

111 A radioisotope is called a tracer when it is used to
   1 kill bacteria in food
   2 kill cancerous tissue
   3 determine the age of animal skeletal remains
   4 determine the way in which a chemical reaction occurs

If you choose this group, be sure to answer questions **112–116.**

112 Which measurement contains a total of three significant figures?
   (1) 0.012 g   (3) 1,205 g
   (2) 0.125 g   (4) 12,050 g

113 The diagram below shows a section of a 100-milliliter graduated cylinder.

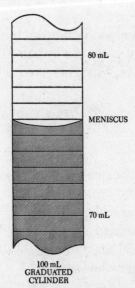

When the meniscus is read to the correct number of significant figures, the volume of water in the cylinder would be recorded as
   (1) 75.7 mL   (3) 84.3 mL
   (2) 75.70 mL   (4) 84.30 mL

114 A student collected the data shown below to determine experimentally the density of distilled water.

Mass of graduated cylinder + distilled
    H_2O sample .... 163 g
Mass of empty graduated cylinder .... 141 g
Mass of distilled H_2O sample ....   g
Volume of distilled H_2O sample .... 25.3 mL

Based on the experimental data collected, what is the density of the distilled water?
(1) 1.0 g/mL   (3) 0.87 g/mL
(2) 0.253 g/mL   (4) 1.15 g/mL

115 A laboratory experiment was performed to determine the percent by mass of water in a hydrate. The accepted value is 36.0% water. Which observed value has an error of 5.00%?
(1) 31.60% water   (3) 37.8% water
(2) 36.0% water   (4) 41.0% water

116 Which is the safest technique for diluting concentrated sulfuric acid?
1 Add the water to the acid quickly.
2 Add the water to the acid and shake rapidly.
3 Add water to the acid while stirring steadily.
4 Add acid to the water while stirring steadily.

ANSWER KEY
JUNE 1992

PART 1

1. 3	14. 1	27. 2	40. 4	53. 1
2. 4	15. 2	28. 4	41. 1	54. 3
3. 4	16. 4	29. 4	42. 4	55. 1
4. 3	17. 2	30. 2	43. 2	56. 1
5. 3	18. 1	31. 1	44. 1	
6. 1	19. 3	32. 3	45. 4	
7. 2	20. 1	33. 2	46. 3	
8. 1	21. 3	34. 2	47. 2	
9. 3	22. 1	35. 3	48. 3	
10. 2	23. 4	36. 1	49. 3	
11. 4	24. 1	37. 4	50. 2	
12. 3	25. 3	38. 2	51. 2	
13. 3	26. 1	39. 2	52. 3	

PART 2

57. 4	69. 1	81. 2	93. 4	105. 1
58. 3	70. 4	82. 4	94. 3	106. 2
59. 2	71. 1	83. 3	95. 2	107. 4
60. 4	72. 2	84. 1	96. 3	108. 2
61. 2	73. 3	85. 4	97. 4	109. 3
62. 2	74. 2	86. 3	98. 3	110. 4
63. 4	75. 1	87. 1	99. 2	111. 4
64. 3	76. 2	88. 3	100. 2	112. 2
65. 2	77. 4	89. 2	101. 1	113. 1
66. 1	78. 3	90. 1	102. 2	114. 3
67. 1	79. 2	91. 4	103. 3	115. 3
68. 3	80. 1	92. 3	104. 2	116. 4

ANSWERS AND EXPLANATIONS
JUNE 1992

PART 1

1. **3** Chemical changes only take place if a compound is present. Elements do not undergo chemical changes. Of the four choices given, methanol (choice 3) is the only compound and is, therefore, the correct answer. Beryllium, boron, and magnesium are all elements.

2. **4** Crystalline solids are the most ordered phase of matter. Solids differ from liquid and gas because a high degree of order and low energy is associated with them. The particles that make up a crystalline solid are close to each other, and they are highly symmetrical in their placement inside the crystal. The answer that most accurately states the arrangement of particles in crystals is choice 4.

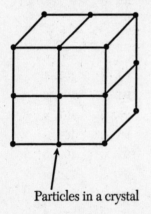

Particles in a crystal

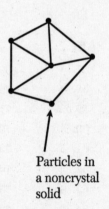

Particles in a noncrystal solid

3. **4** The energy associated with the heat of vaporization is required to break intermolecular bonds during the vaporization process. The stronger and greater the number of intermolecular bonds present in solution, the greater the amount of energy needed to break them. In other words, solutions with the strongest and highest number of intermolecular bonds have the greatest heat of vaporization. Based on this

observation, choice 4 must be the right answer because it has the largest heat of vaporization.

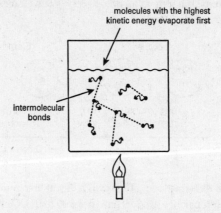

molecules with the highest kinetic energy evaporate first

intermolecular bonds

4. 3 Use Reference Table O to answer this question. The vapor pressure and the boiling point of a liquid are directly proportional to each other. As one increases, so does the other. We also know that liquids boil when the vapor pressure is equal to the atmospheric pressure. According to Table O, water has a vapor pressure of 355.1 torr at 80°C. Therefore, if the atmospheric pressure is 355.1 torr as well, water boils at 80°C.

5. 3 There are two types of reactions when it comes to thermodynamics. One type is called an *endothermic reaction*, and the other type is known as an *exothermic reaction*. During an endothermic reaction, energy is absorbed from the environment to drive the reaction. Thus, the external temperature is lowered. On the other hand, exothermic reactions release energy into the environment, increasing the temperature. Because the temperature of the water is increased, the reaction that dissolved the substance in water must have been exothermic.

6. 1 The center of an atom is made up of the nucleus. The nucleus occupies a very small region within an atom. Although the nucleus is small, the majority of the nuclear mass is concentrated within the nucleus. Protons and neutrons, which are located inside the nucleus, account for most of an atom's mass. Electrons have $\frac{1}{1836}$ the mass of a proton. Choice 1 correctly states the physical characteristics of an atom.

7. **2** Isotopes, by definition, have the same atomic number (same number of protons) but different atomic masses (different number of neutrons). Therefore, all isotopes of a given element must have the same atomic number.

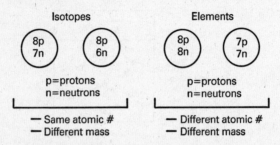

Isotopes	Elements
8p 7n 8p 6n	8p 8n 7p 7n
p=protons n=neutrons	p=protons n=neutrons
— Same atomic # — Different mass	— Different atomic # — Different mass

8. **1** In the equation $^{234}_{90}\text{Th} \rightarrow {}^{234}_{91}\text{Pa}$, the atomic number (the number of protons) increased by one (from 90 to 91). Only one type of particle can increase the atomic number by one—a beta particle. A beta particle, or $_{-1}^{0}\text{e}$, is correctly represented in choice 1.

$$\text{Th}_{90 - (-1 \text{ the charge of beta particle})} = \text{Pa}_{91[90-(-1)=91]}$$

In a balanced nuclear equation the sum of the mass numbers on the left side of the equation must equal the sum of the mass numbers on the right. Similarly, the sum of the atomic numbers on the left must equal the sum of the mass numbers on the right. Because there is no change in the mass number from Th to Pa, the mass number of the missing particle must be zero. Because the atomic number is increased by one, the atomic number of the missing particle must be negative one.

9. **3** All four valence electrons are in the ground state in subshells s and p. The s subshell can hold up to two electrons. The p subshell can hold up to six electrons. If there are four valence electrons, two electrons are found in the s subshell and two electrons are found in the p subshell. Therefore, the correct answer is choice 3.

Orbital diagram

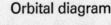

s	p_x	p_y	p_z
↑↓	↑	↑	

10. 2 One atomic mass unit is approximately equal to the mass of a proton. In addition, protons are positively charged particles that are found in a nucleus of an atom. The answer can be also found in Reference Table J.

11. 4 In the atom carbon-14, the number 14 is the atomic mass number. The atomic mass number is the sum of all the protons and neutrons in a nucleus. A neutral atom has the same number of electrons and protons but can have a different number of neutrons. The correct answer is choice 4 because carbon-14 has six protons and eight neutrons (6 + 8 = 14) and six electrons.

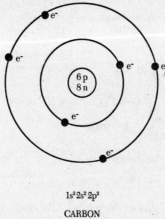

$1s^2 2s^2 2p^2$

CARBON

12. 3 In the electron configuration $1s^2 2s^2 2p^3$, the highest energy sublevel is $2p$. It has three electrons in it. Therefore, the correct answer is choice 3.

13. 3 If you draw an electron dot formula, you will see how the valence electrons are shared between two atoms. H has one electron and Cl has seven electrons in their respective valence shells. Based on the formula, the correct answer is choice 3.

14. 1 In 1.0 mole of $Al_2(CrO_4)_3$, there are 12 moles of oxygen atoms (4×3). One mole of oxygen atoms weighs 16 grams. Therefore, 12 atoms of oxygen have a total mass of $16 \text{ g} \times 12 = 192 \text{ g}$.

15. 2 Choices 1 (HCl), 3 (H_2O), and 4 (NH_3) are all polar molecules because of the great electronegativity difference between the atoms of each one of the molecules and the lack of symmetry in the molecules.

As a result, a partial charge is developed over each one of the molecules. Although there is a great electronegative difference between carbon and hydrogen atoms of CH_4, they are symmetrically arranged and cancel out each other's partial charges (see diagram). As a result, the molecule has no overall charge. In other words, the molecule is nonpolar.

Symmetric Shape
METHANE

Asymmetric Shape
WATER

Tetrahedral
Shape

16. 4 Ammonium (NH_4) has a charge of +1. Carbonate, or CO_3, has a charge of –2. Therefore, the correct formula for ammonium carbonate is $(NH_4)_2CO_3$.

17. 2 Network solids are held together by covalent bonds formed between neighboring atoms. Therefore, the correct answer is choice 2. Hydrogen bonds are intermolecular bonds and occur between molecules containing H atoms. Ionic bonds occur between metals and nonmetals. Metallic bonds are present in metals.

18. 1 The most stable bond has the least amount of energy associated with it. In other words, when the bond is formed a large quantity of energy is released. In the four choices given, when the H—F bond was formed the greatest amount of energy was released (135 kcal/mole). Therefore, it must be the most stable bond.

19. 3 The groups are listed from top to bottom (vertically) in the periodic table, and the periods are arranged from left to right (horizontally).

According to the periodic table, radium (Ra) lies in group 2, period 7. Magnesium (Mg) is located in group 2, period 3; manganese (Mn) is located in group 7, period 4; and radon (Rd) is located in group 18, period 6.

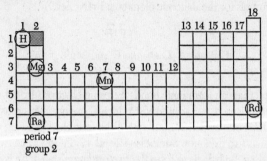

period 7
group 2

20. 1 Consult Reference Table P to answer this question. According to Table P, iodine (I) has the largest radius. Therefore, its ion, I⁻, also has the largest atomic radius. The negative ion of an atom is larger than the neutral atom.

Atom	Covalent radius (Å)
F	0.64
Cl	0.99
Br	1.14
I	1.33

21. 3 An atom in the ground state that contains eight valence electrons has all its s and p subshells filled. Its electron configuration resembles ns^2np^6. From the periodic table, we can determine that the group of elements with this electron configuration is the noble gases.

22. 1 Only choice 1 is a true characteristic of a nonmetal.

Common Characteristics of Metals and Nonmetals

Metals	Nonmetals
Malleable	Brittle
Shiny luster	No luster
Good electrical conductor	Poor electrical conductor
Good heat conductor	Poor heat conductor

23. **4** Halogen group elements F_2, Cl_2, Br_2, and I_2 are diatomic molecules at STP. Halogens belong to group 17 (VIIA). Diatomic molecules form nonpolar covalent bonds. Another easy way to remember the diatomic elements is the name Brinclhof, which can be spelled with the symbols for the diatomic elements: BrINClHOF.

$$:\ddot{F}:\qquad\qquad:\ddot{F}:\ddot{F}:$$

F atom   Diatomic F_2 molecule

24. **1** A metal that can form ionic bonds by losing electrons from both the outermost and next-to-outermost principal energy levels must have more than one oxidation number. Of the four choices given, only Fe (iron) has two oxidation numbers: +2 and +3. Fe loses electrons from $4s^2$ and $3d^{10}$ subshells to form the two oxidation states. Fe is also the only transition metal listed.

25. **3** The easiest way to solve this problem is to determine in which compound the ratio of hydrogen to oxygen is the highest. The molecule that has the highest H:O ratio also has the greatest percent by mass of hydrogen. Based on the ratios below, the greatest percent by mass of hydrogen is present in H_3O^+ (choice 3).

Compound	Ratio (H:O)
OH^-	1:1
H_2O	2:1
H_3O^+	3:1
H_2O_2	2:2, or 1:1 when reduced to lowest possible coefficients

26. **1** In the balanced reaction $CH_4 + 2O_2 \rightarrow CO_2 + 2H_2O$, for every 1 mole of CH_4 that reacts, 2 moles of O_2 are consumed. In other words, CH_4 and O_2 react with each other in a molar ratio of 1:2. Therefore, if 3.0 moles of CH_4 completely reacts with oxygen, the following number of moles of oxygen must be present:

$$\frac{1}{3} = \frac{2}{x}$$

x = 6 moles of oxygen

27. 2   In 1.0 molecule of nitrogen (N_2) there are two nitrogen (N) atoms. Therefore, in 1.0 mole of N_2 molecules there are 2.0 moles of N atoms. Based on this ratio, in 2.0 moles of N_2, there are 4.0 moles of N atoms.

28. 4   One mole of any element contains 6.0×10^{23} atoms. Therefore, the correct answer is the choice that states the molar mass of an element. According to the periodic table, 1 mole of silicon has a mass of 28 grams. Therefore, the correct answer is choice 4.

29. 4   A mixture contains at least two different molecules or ions. The only choice where more than one compound is present is in an aqueous HCl solution. H^+ and Cl^- ions are present in an aqueous HCl solution. In choice 1, iodine is in solid phase; in choice 2, iodine is in liquid phase; and in choice 3, HCl is in gaseous phase. Ions do not exist in solid, liquid, or gaseous phases. Only aqueous solutions are homogenous mixtures because they contain two or more ions and molecules of water.

30. 2   All catalysts lower the activation energy of a reaction. When activation energy is lowered, the reaction rate increases. Catalysts affect no other aspects of reactions. Choice 2 correctly states the function of a catalyst.

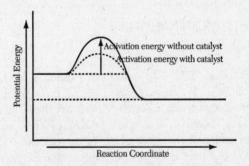

31. 1   The function of Bunsen burners is simple. They are used to supply energy in the form of heat to reactants. Reactions usually require energy to start themselves (activation energy). According to choice 1, the reactants are activated by heat from the Bunsen burner flame. Based on our knowledge of heat and reactions, choice 1 is the correct answer.

32. 3   The reaction rate is determined by two factors: the frequency of collisions between the reactants and the effectiveness of those collisions. Increasing the temperature of reactants influences both factors. Raising the temperature increases the average kinetic energy of reactants. With greater kinetic energy, reactants collide more often and with greater force.

33. 2   Most spontaneous reactions are also exothermic reactions. In an exothermic reaction, energy is released. In other words, the enthalpy decreases. An increase in randomness, or entropy, also favors a spontaneous reaction. Remember, the free energy of spontaneous reactions must be negative.

$$\Delta G \qquad = \Delta H \quad - \quad T\Delta S$$

free energy   enthalpy   entropy

34. 2   Salts are composed of a metal and a nonmetal. Only choice 2 (KCl) is a salt. It is made of the metal potassium (K) and the nonmetal chlorine (Cl). Choice 1 (KOH) is a base, choice 3 (CH_2OH) is an alcohol, and choice 4 (COH_3COOH) is an acid.

35. 3   According to the ionization constant of water, the product of the concentrations of H^+ and OH^- ions always equals 1.0×10^{14}. Therefore, the hydronium concentration $[H_3O^+]$ is

$[H^+][OH^-] = 1.0 \times 10^{-14}$

$[H_3O^+][1.0 \times 10^{-3}] = 1.0 \times 10^{-14}$

$[H_3O^+] = 1.0 \times 10^{-11}$.

36. 1   Use Reference Table L to answer this question. Table L lists the acid dissociation, or K_a, values of various acids. The higher the K_a value, the stronger an acid is. Conversely, the lower the K_a value, the weaker an acid is. Therefore, by comparing the K_as of a group of acids, we can determine whether they are strong or weak acids. According to Table L, H_2S has the lowest K_a of the four choices given. The K_a of H_2S is 9.5×10^{-8}. Therefore H_2S is the weakest acid.

37. 4   A Brönsted-Lowry acid can donate a proton (H^+). In the given reaction

$NH_4^+ + OH^- \rightarrow H_2O + NH_3$,

NH_4^+ is the Brönsted-Lowry acid. It donates a proton to the electron-rich OH^-. H_2O is the Brönsted-Lowry acid in the reverse reaction.

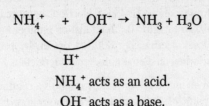

$$NH_4^+ \;+\; OH^- \rightarrow NH_3 + H_2O$$

$$H^+$$

NH_4^+ acts as an acid.

OH^- acts as a base.

38. 2 According to the reaction

$$H_2SO_4 + 2NaOH \rightarrow Na_2SO_4 + 2H_2O,$$

2 moles of NaOH are required to neutralize 1 mole of H_2SO_4. Because there is only 1 mole present in 1 liter of NaOH (1 mol = 1 mol/L), 2 liters of 1 M NaOH provides the 2 moles of NaOH that are necessary to neutralize the acid.

39. 2 NH_2^- is the conjugate base of NH_3 because the two species are related by the donation of a proton (H^+). When NH_3 loses a proton, it becomes NH_2^-.

40. 4 A compound that turns a red litmus blue, conducts electricity, and reacts with an acid to form a salt and water must have the following characteristics: It must be ionic in nature and a strong base. Of the four choices given, only choice 4 (LiOH) satisfies all the characteristics mentioned above. Choice 1 (HCl) is an acid. Choices 2 (NaI) and 3 (KNO_3) are salts.

41. 1 Oxygen has a positive oxidation number if it reacts with an element whose electronegativity is higher than that of oxygen's. In a reaction with an element with higher electronegativity, oxygen is forced to oxidize. Of the four elements given, only fluorine (F) has a higher electronegativity.

Element	Electronegative value
F	4.0
O	3.5
Cl	3.2
Br	2.9
I	2.7

42. 4 In the given reaction, Al(s) loses electrons because its oxidation number changes from 0 to +3. The oxidation number increases when an element loses electrons or oxidizes. Conversely, the oxidation number is lowered when an element undergoes reduction.

43. 2 An electrochemical cell or battery requires a continuous flow of charged particles. To maintain a continuous flow of charged particles, a complete circuit must be present. A salt bridge is necessary to form a complete circuit. A salt bridge allows ions to migrate from one half-cell to the other (refer to the diagram). The answer that accurately describes the function of a salt bridge is choice 2.

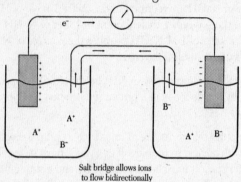

Salt bridge allows ions
to flow bidirectionally

44. 1 In the given reaction, Pb + 2Ag$^+$ → Pb^{2+} + 2Ag, Ag$^+$ gained an electron because its oxidation number changed from +1 to 0. The oxidation number is lowered when an element undergoes reduction. On the other hand, the oxidation number increases when an element loses electrons or oxidizes.

45. 4 Oxidation is the process through which an element loses electrons from its valence shell. The oxidation number increases when an element undergoes oxidization. The only choice that accurately depicts oxidation is the reaction in choice 4, in which Mg loses two electrons to become Mg^{2+}. In choices 1 and 2, electrons are being added to Mg instead of subtracted. In choice 3, the oxidation number decreases instead of increasing in the resulting Mg ion.

46. 3 For an equation to be balanced, the same numbers of each atom must appear on both sides of the reaction, and the total charges must

be balanced as well (electrons gained = electrons lost). By trial and error, the following is the balanced equation:

$$3Pb^{2+} + 2Au^{3+} \rightarrow 3Pb^{4+} + 2Au$$

Based on this balanced equation, Pb^{+2} has 3 as its smallest whole-number coefficient.

Another way to balance redox reactions is to first write out the half reactions.

$$Pb^{2+} \rightarrow 2e^- + Pb^{4+}$$

$$Au^{3+} + 3e^- \rightarrow Au$$

Multiply the first equation by 3 and the second by 2, so that six electrons are lost and six electrons are gained. Now you have the smallest whole number ratios for the balanced equation.

47. **2** The alcohol that reacts with C_2H_5COOH to produce the ester $C_2H_5COOC_2H_5$ is C_2H_5COOH (see the reaction below).

Condensation reaction

48. **3** A compound with the molecular formula C_7H_8 has the general formula C_nH_{2n-6} where $n = 7$. Therefore, $C_7H_{[2(7)-6]} = 8$. In all other choices, C_7H_8 does not fit the indicated general formula.

49. **3** Fermentation is part of anaerobic respiration. During fermentation, sugar is oxidized to produce CO_2 and ethanol (alcohol). Glucose and zymase are necessary to carry out fermentation, but they are not products of fermentation. Esterification produces esters (R_1COOR_2). Saponification produces soap (an organic molecule with polar and nonpolar ends). Polymerization is the general reaction through which small molecules are bound to produce a large polymer or molecule.

50. 2 Isomers are compounds with identical molecular formulas but different structural formulas. The isomer of the given structure is accurately depicted in choice 2 (see diagram). While the molecules are structurally different from each other, they have the same number of carbons (3), hydrogens (6), and oxygen (1) in their structure

$$
\begin{array}{ccc}
\text{H} & \text{H} & \text{H} \\
| & | & | \\
\text{H}-\text{C}-\text{C}-\text{C}=\text{O} \\
| & | \\
\text{H} & \text{H}
\end{array}
\qquad
\begin{array}{ccc}
\text{H} & \text{O} & \text{H} \\
| & \| & | \\
\text{H}-\text{C}-\text{C}-\text{C}-\text{H} \\
| & & | \\
\text{H} & & \text{H}
\end{array}
$$

# of carbons = 3   # of carbons = 3
# of hydrogens = 6   # of hydrogens = 6
# of oxygen = 1   # of oxygen = 1
Molecular formula = C_3H_6O Molecular formula = C_3H_6O
Structural formula = C_2H_5CHO Structural formula = CH_3COCH_3
(Aldehyde)   (Ketone)

51. 2 An addition reaction can only take place in an unsaturated molecule. Unsaturated molecules include alkenes and alkanes. Of the four choices given, C_2H_4 is an alkene, and CH_4, C_3H_8, and C_4H_{10} are all alkanes. Alkenes are characterized by double bonds. Below is an example of an alkene molecule undergoing addition reaction.

$$
\begin{array}{cc}
\text{H} & \text{H} \\
| & | \\
\text{H}-\text{C}=\text{C}-\text{H}
\end{array}
+ Cl_2 \rightarrow
\begin{array}{cc}
\text{H} & \text{H} \\
| & | \\
\text{H}-\text{C}-\text{C}-\text{H} \\
| & | \\
\text{Cl} & \text{Cl}
\end{array}
$$

unsaturated
hydrocarbon
(alkene)

52. 3 In a closed system, an equilibrium between the amount of vapor formed and the amount of liquid in the container is reached over time. Because it is a closed system, vapors cannot escape. As a result, as vapor molecules form, some of them condense and revert to liquid phase at the same rate so that there is no overall (macroscopic) change in the concentration of the vapor. In an open system, the rate of vaporization is greater than the rate at which liquid forms.

53. 1 According to Boyle's law, as pressure increases, volume decreases, and vice versa, at constant temperature. In other words, the pressure and volume are inversely related to each other. Therefore, if the pressure of the gas increases, the volume of the gas decreases.

54. 3 Pressure only affects molecules and elements that are in their gaseous phase. All other phases are unaffected by pressure. When pressure increases, the reaction shifts from the side with the higher number of moles of gases to the side with the lower number of moles of gases. Conversely, when pressure is lowered the reaction shifts to the side with the higher number of moles of gases. Lastly, if the number of moles is the same on both sides then the pressure has no effect on the reaction. Because the number of moles of gases is the same on both sides (2 moles), the number of moles of NO remains the same when pressure increases.

55. 1 When chemical bonds are formed, energy is released. Release of energy makes the newly formed compound or diatoms more stable than the original reactants, so the potential energy of the original atoms is decreased.

56. 1 As we move from left to right (atomic number of elements increase in the same direction), electronegativity increases. Elements with higher electronegativity tend to reduce (gain electrons) and not oxidize themselves. An element that oxidizes itself acts as a reducing agent. Therefore, as we move from left to right, the ability to act as a reducing agent decreases with increasing electronegativity.

PART 2
Group 1—Matter and Energy

57. 4 Temperature represents the average kinetic energy of all molecules of a compound in a given phase. Therefore, an increase in temperature indicates an increase in the average kinetic energy of molecules. In the problem, the temperature increases from 273 K (0°C) to 298 K. This means that the average kinetic energy of the water molecules increased. The number of water molecules remains the same, however, because each sample consists of a mole of water molecules. The correct answer is choice 4 because it accurately describes the relationship between the two samples.

58. 3 A binary compound is composed of two elements. The prefix *bi-* means two. Of the four answer choices given, only choice 3 (C_3H_8) is a binary compound. It is made of carbon and hydrogen. Be careful

not to confuse binary with diatomic. Diatomic species contain only two atoms that can be of the same element as in choice 2.

59. **2** The element chlorine undergoes a change of phase when the temperature is decreased from 25°C. According to Reference Table C, the boiling point of chlorine is 238 K, or –35°C. Therefore, if the temperature of chlorine is reduced from 25°C to –35°C, the chlorine changes from gaseous to liquid phase. Aluminum, silicon, and sulfur are all solids at 25°C, so they do not change phase even if the temperature is lowered.

60. **4** According to the diagram, as a substance changes from a solid to a liquid to a gas, energy is absorbed. Conversely, when substances change from a gaseous to a liquid to a solid phase, energy is released. Only choice 4 indicates a release in energy. Condensation is the process in which gases change to liquid. Therefore, choice 4 is the correct answer. All other choices require absorption of energy.

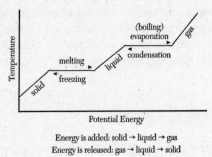

Energy is added: solid → liquid → gas
Energy is released: gas → liquid → solid

61. **2** The partial pressure of a gas within a mixture is calculated by multiplying the gas's mole ratio with the total pressure of the mixture.

$$\text{Mole ratio of argon} = \frac{\text{moles of argon}}{\text{total moles in mixture}}$$

$$= \frac{3.0 \text{ moles}}{(2.0 \text{ moles} + 3.0 \text{ moles} + 5.0 \text{ moles})}$$

$$= 0.3$$

At STP, the gas mixture has pressure of 1 atm or 760 torr. Therefore, the partial pressure of argon = 760 torr × 0.3 = 228 torr.

Group 2—Atomic Structure

62. **2** Each sublevel contains a specific number of orbitals. The number of orbitals determines the total number of electrons that are found in each sublevel. The total number of orbitals in sublevel d is 5.

Sublevel	Number of orbitals present
s	1
p	3
d	5
f	7

63. **4** Consult Reference Table H to answer this question. Table H lists various isotopes, their half-lives, and their mode of decay. According to Table H, ^{37}K has a half-life of 1.23 seconds and decays by emitting a positron (β^+). On the other hand, ^{42}K has a half-life of 12.4 hours and decays by emitting an electron (β^-). Therefore, ^{42}K has a longer half-life than ^{37}K, and they differ in the mode in which they decay.

64. **3** According to Reference Table H, the half-life of ^{131}I is 8.07 days.

Time elapsed	Mass ^{131}I remaining
8 days	½ of original mass
16 days	¼ of original mass
24 days	⅛ of original mass
32 days	¹⁄₁₆ of original mass
40 days	¹⁄₃₂ of original mass

According to the table above, ^{131}I undergoes five half-lives. Therefore, if 1.0 gram of ^{131}I is left after 40 days, then original mass must have been
$x/32 = 1.0$ g
$x = 32$ g.

65. 2 Chlorine has an atomic number of 17. Therefore, chlorine must have a total of 17 protons and 17 electrons. Based on the fact that chlorine has 17 electrons, the electron configuration must be $1s^22s^22p^63s^23p^5$. The configuration shows that an atom of chlorine in the ground state has five sublevels: $1s$, $2s$, $2p$, $3s$, and $3p$.

66. 1 When an electron absorbs energy, it is elevated from a lower energy level to a higher energy level. An elevated electron is better known as an *excited electron*. When excited electrons fall from a higher energy level to a lower energy level, energy is given off, producing a bright-line spectrum of light. Energy is released only when electrons fall to a lower energy subshell.

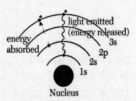

"Falling" electrons emit light

Group 3—Bonding

67. 1 To conduct electricity, charged particles must be present. If the charged particles are bonded to each other in a compound, then the compound is not able to conduct electricity. Charged particles only conduct electricity if they are dissolved in solution and separated from each other. The only solid capable of conducting electricity when dissolved in water is an ionic solid. An ionic solid or compound dissolves in water to produce positive and negative charges. The resulting aqueous solution is capable of conducting electricity. In the problem, it is noted that the substance has a melting point of 1,074 K. This is an extremely high melting point. Ionic compounds usually have a very high melting point.

68. 3 The symbol Ne represents a mole of neon, a gram of Ne is represented as 1 g Ne, and a liter of Ne is represented as 1 liter of Ne. One atomic mass unit of neon makes no sense because a single proton is 1 atomic mass unit, and a single atom of neon has 10 protons and 10 neutrons.

69. 1   Water molecules are polar in nature. The two components of water
         (hydrogen and oxygen) are bonded to each other through covalent
         bonds. Because hydrogen and oxygen have great differences in elec-
         tronegativity, the covalent bond is polar. The molecule is polar
         because of asymmetry and the two lone pairs of oxygen electrons.
         Based on these characteristics of the bond, the correct answer is
         choice 1.

70. 4   According to the diagram in the exam, when charged particles or ions
         are dissolved in water, they are surrounded by water molecules.
         Therefore, it is reasonable to conclude that there is an attraction
         between ions and water molecules. In other words, the dissolving
         process takes place by molecule-ion attractions.

71. 1   This equation can only be balanced by the trial-and-error method.
         The balanced equation is $Al_2(SO_4)_3 + 3ZnCl_2 \rightarrow 2AlCl_3 + 3ZnSO_4$.
         According to the balanced equation, the sum of the coefficient is
         $1 + 3 + 2 + 3 = 9$.

Group 4—Periodic Table

72. 2   Salts of transition metal ions form colorful solutions. Of the four
         answer choices given, only $CuSO_4$ is a salt of a transition element.
         Transition elements are present between groups 3 and 11 in the
         periodic table.

73. 3   Use the periodic table to answer this question. The halogens, located
         in group 17, have elements that show oxidation states that are both
         positive and negative. For example, fluorine (F) has an oxidation
         number of -1. On the other hand, chlorine (Cl) has oxidation states of
         $-1, +1, +3$, and so forth. Transition metals have a variety of oxidation
         states, but they are all positive.

74. 2   According to Reference Table P, as the atomic number or nuclear
         charge increases from left to right in a period, the covalent atomic
         radius decreases or gets smaller. Therefore, choice 2 is the correct
         answer. Covalent atomic radius decreases from left to right because
         effective nuclear charge increases from left to right.

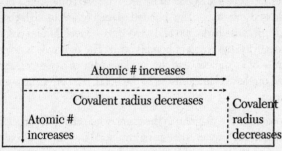

Atomic # increases

Covalent radius decreases

Atomic # increases

Covalent radius decreases

Periodic Table

75. 1 Potassium (K) belongs to the alkali group (group 1) in the periodic table. Group 1 metals are very reactive and are found only in compounds in nature. Pure potassium tends to oxidize very quickly and combines with other elements to form compounds.

76. 2 Iodine is a solid at STP. Bromine and mercury are liquids at STP, and neon is an inert gas.

Group 5—Mathematics of Chemistry

77. 4 Use the following equation to solve the problem: heat (in calories) = mass of water × change in temperature × specific heat of H_2O. By definition, one calorie is equal to the heat required to raise 1 g of water 1°C, so the specific heat of water is 1 calorie/g°C.

$$= 100 \text{ g} \times (20°C - 16°C) \times \text{calorie/g°C}$$

$$= 100 \text{ g} \times 4°C \times 1 \text{ calorie/g°C}$$

$$= 400 \text{ g°C} \times 1 \text{ calorie/g°C}$$

$$= 400 \text{ calories}$$

78. 3 When a solute (any particle) is added to a pure liquid, the boiling point of the solution as compared to the solvent is increased and the freezing point is decreased. These properties of solutions are called the *colligative properties*. The extent of these changes depends on the concentration of the solute dissolved. Boiling point is increased because it is more difficult for solvent particles to escape into the

vapor phase, and the freezing point is lowered because it is more difficult for the solvent particles to arrange themselves in a lattice with the addition of a solute.

79. 2 Divide the mass of carbon and oxygen by their respective molar masses to obtain the number of moles.

$$C = \frac{24 \text{ g}}{12 \text{ g}} = 2.0 \text{ mol}$$

$$O = \frac{64 \text{ g}}{16 \text{ g}} = 4.0 \text{ mol}$$

Next, use the calculated moles to form a formula. Mole ratios are used as subscripts for the formula: C_2O_4. Finally, reduce the subscripts to the lowest whole-number ratio to obtain the empirical formula. Based on the mole ratio, the empirical formula must be CO_2.

80. 1 Calculate the density of C_2H_6.

Density = mass/volume

Mass of 1 mol of C_2H_6 = 24 g (carbon 12×2) + 6 g (hydrogen 1×6) = 30 g

Volume of 1 mol of C_2H_6 at STP = 22.4 L

Therefore, density is equal to D = 30 g/22.4 L = 1.3 g/L

Refer to Reference Table C to find out which gas has a density similar to C_2H_6. According to Table C, nitrogen oxide (NO) has a density of 1.34 g/L. Because 1 mole of all gases at STP occupies the same volume (22.4 L), the gas that has the same density (g/L) as C_2H_6 has the same molar mass. The molar mass of C_2H_6 is 30 g/mol.

2×12 g/mol + 6×1 g/mol = 30 g/mol

Gas	Molar mass
NO	30 g/mol
NH_3	17 g/mol
H_2S	34 g/mol
SO_2	64 g/mol

NO has the same molar mass as C_2H_6 and therefore the same density at STP.

81 2 Use the combined gas law formula to solve this problem.

$$\frac{P_1 V_1}{T_1} = \frac{P_2 V_2}{T_2}$$

$$T_2 = \frac{P_2 V_2 T_1}{P_1 V_1}$$

$$= \frac{273 \text{ K} \times 380 \text{ mm Hg} \times 551 \text{ ml}}{760 \text{ mm Hg} \times 400 \text{ ml}} = 188 \text{ K}$$

Group 6—Kinetics and Equilibrium

82. 4 The salt that has the lowest K_{sp} (solubility constant) is the least soluble of the four salts given in the answer choices. K_{sp} tells us the extent to which any compound dissolves in water. The higher the K_{sp}, the greater the solubility. Conversely, the lower the K_{sp}, the lower the solubility. According to Reference Table E, AgBr has the smallest K_{sp} because it is nearly insoluble. $AlBr_3$ is soluble, $PbBr_2$ is slightly soluble, and NaBr is soluble.

83. 3 K_{sp} is the product of the products formed in the dissociation reaction.

$$AB \rightarrow A^+ + B^-$$

$$K_{sp} = [A^+][B^-]$$

The concentrations of the ions are raised to the power of their coefficients in the balanced dissociation reaction. Therefore, the K_{sp} of the reaction in the question is $K_{sp} = [Ca^{2+}][F^-]^2$.

84. 1 All reactions that are spontaneous in nature possess a negative free energy ($-\Delta G$). According to Reference Table G, only CO_2 has a negative free energy of formation (-94.3 kcal/mol). All other choices have positive free energy ($+\Delta G$).

85. 4 The reaction $2Na + 2H_2O \rightarrow 2Na^+ + 2OH^- + H_2$ reaches completion if solid sodium crystals (Na solid), water, and a gas are formed. The formation of any one of the three molecules prevents the reverse reaction from taking place. Of the three factors that bring this reaction to a completion, only gas is an answer choice.

86. 3 When KCl is added to a solution, the concentration of Cl⁻
increases. This increase in Cl⁻ ion concentration upsets the equilibrium, and, according to the Le Châtelier principle, the system must
readjust to re-establish equilibrium. Some of the added Cl⁻ reacts
with Pb^{2+} to form $PbCl_2$ until equilibrium is reached. This change
is a shift to the left, and the concentration of Pb^{2+} decreases as a
result. The answer that accurately describes the stress on the given
reaction is choice 3.

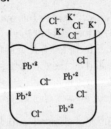

When KCl is added, the [Cl⁻] increases.
As a result, the reaction is shifted to the left.

Group 7—Acids and Bases

87. 1 K_a, or acid dissociation, is represented as the equilibrium concentration of the products divided by the equilibrium concentration of the
reactants.

$$K_a = \frac{[H^+]\,[\text{conjugate base}]}{[\text{unionized acid}]}$$

Therefore, the K_a of the acid HF is

$$K_a = \frac{[H^+]\,[F^-]}{[HF]}\;.$$

88. 3 An amphoteric substance can act as an acid or base because it can
both donate a proton as well as accept a proton. Therefore, based on
the definition of what an amphoteric substance is, all amphoteric substances (with the exception of H_2O) must have a hydrogen and a negative charge (or readily available electrons) as part of their structure.
The hydrogen indicates a proton, and the negative sign (or a lone pair

of electrons) indicates the substance's ability to accept a proton (which is positively charged). Choice 3 (HCO_3^-) is the only answer choice that possesses both a hydrogen and a negative charge. ($HCO_3^- + H_2O \rightarrow H_2CO_3 + OH^-$ acts as a base and $HCO_3^- + H_2O \rightarrow CO_3^{2-} + H_3O^+$ acts as an acid.)

89. 2 An Arrhenius acid must be able to donate a proton or an ion of hydrogen. Of the choices given, only HI is found in Reference Table L.

90. 1 Use Reference Table N to answer this question. The most active metals are listed on the bottom right of the half-reactions listed in Table N. Some active metals include Li, K, and Cs. On the other hand, less active metals are located on the top right side of Table N. Some less active metals include Ag, Cu, and Pb. Based on Table N, the most active metal of the four answer choices given is aluminum.

91. 4 The greater the number of charged particles in the solution, the brighter the bulb glows. To answer this question, both Reference Tables L and M must be consulted. The substance with the highest dissociation constant produces the greatest number of ions (charged particles) in solution. As a result, in the presence of that substance, the bulb glows the brightest. Of the four answer choices given, sulfuric acid (H_2SO_4) has the highest dissociation constant, or K_s.

Group 8—Redox and Electrochemistry

92. 3 Determine the species that undergoes reduction and the species that undergoes oxidation in the given reaction: Br_2 undergoes reduction and Zn undergoes oxidation. Next, determine the electrode potentials of the Zn and Mg from Reference Table N.

Zn = +0.76 V (The sign of the reduction potential is switched because zinc is being oxidized in this reaction.)

Br_2 = +1.09 V

Finally, find the cell voltage (E^0) by adding the electrode potential of the species being oxidized and that of the species that is being reduced.

$$E^0 = E^0_{reduced} + E^0_{oxidized}$$
$$= (+0.76 \text{ V}) + (+1.09 \text{ V}) = +1.85 \text{ V}$$

93. 4 According to Reference Table N, the half-reaction $2H^+ + 2e^- \rightarrow H_2$ has a reduction potential of 0.00 V. Because H^+ reduction potential is zero, it is arbitrarily used as the standard for electrode potentials.

94. 3 Any element that reacts spontaneously with Al^{+3} must have a lower reduction potential than Al's reduction potential. Of the choices given, only Li lies below Al on Reference Table N.

$3Li + Al^{3+} \rightarrow 3Li^+ + Al$

95. 2 The fork is attached to the negative terminal of the battery. That means that the fork has excess electrons passing through it. As a result, it acquires a negative charge. The negatively charged fork acts as a cathode because it attracts Ag^+, which turns into Ag metal on the surface of the fork. Remember that reduction takes place at the cathode.

96. 3 The reaction at the fork is a reduction reaction. Ag^+ gains an electron, and its oxidation number is decreased.

$Ag^+ + e^- \rightarrow Ag$ (reduction)

Group 9—Organic Chemistry

97. 4 Nylon is composed of many diamine and dicarboxylic acids. These two molecules combine with each other to form a large molecule called *nylon*. When small molecules react with each other to form a large molecule, the reaction is termed a *polymerization reaction*.

98. 3 Among hydrocarbons, molecules with the highest molecular mass and the longest chain have the highest boiling point. Large molecules require more energy to change phases because their van der Waals forces are the strongest. Of the four choices, pentene has the highest molecular weight. Therefore, it has the highest boiling point as well.

99. 2 The structure of the 2-propanol must include an alcohol present on the second carbon of the propanol chain, which has three carbons.

2-Propanol

100. 2   Every organic acid must have the following functional group: –COOH (carboxyl group).

$$\underset{RCOH}{\overset{O}{\overset{\|}{}}} \quad \text{Carboxylic Acid}$$

$$\underset{RCH}{\overset{O}{\overset{\|}{}}} \quad \text{Aldehyde}$$

$$R_1-\underset{}{\overset{O}{\overset{\|}{C}}}-R_2 \quad \text{Ketone}$$

$$R_1-O-R_2 \quad \text{Ethers}$$

101. 1   Ethers have the structural formula R_1—O—R_2. The word *diethyl* indicates that the R_1 and R_2 groups are ethyl groups that have a molecular structure of C_2H_5.

$$H-\underset{\underset{H}{|}}{\overset{\overset{H}{|}}{C}}-\underset{\underset{H}{|}}{\overset{\overset{H}{|}}{C}}-O-\underset{\underset{H}{|}}{\overset{\overset{H}{|}}{C}}-\underset{\underset{H}{|}}{\overset{\overset{H}{|}}{C}}-H$$

$$\underbrace{}_{\text{Ethyl}} \qquad \underbrace{}_{\text{Ethyl}}$$

Diethyl Ether

Group 10—Applications of Chemical Principles

102. 2   The contact process is a series of reactions that are carried out to synthesize sulfuric acid from sulfur, oxygen, and water. When sulfur is burned, the first product formed is sulfur dioxide. Recall that burning implies combustion, or reaction with O_2.

$$S + O_2 \rightarrow SO_2$$

In the second and third steps, SO_2 undergoes more reactions to ultimately produce sulfuric acid.

103. 3   The common gas that is found in petroleum is methane (CH_4). Methane is also called *natural gas*.

104. 2   Fractional distillation is the process by which volatile substances are separated from each other based on their boiling points. Fractional

crystallization separates parts of liquid by cooling the mixture; each component of the liquid freezes at its own freezing point to form crystals. Esterification and saponification are separation processes that involve chemical reactions.

105. 1 In the given reaction

$$Pb + PbO_2 + 2H_2SO_4 \rightarrow 2PbSO_4 + 2H_2O$$

Pb^0 is oxidized to Pb^{+2} (Pb has a charge of +2 in $PbSO_4$). Recall that in oxidation an element's oxidation number increases.

106. 2 Carbon is used to make iron (Fe). When iron ore is reacted with carbon and heated, it yields pure iron (Fe).

$$2Fe_2O_3 + 3C \rightarrow 4Fe + 3CO_2$$

Group 11—Nuclear Chemistry

107. 4 Carbon dating is a common method for determining the age of artifacts made of organic material, such as wood. The ratio of carbon-14, which is a radioactive isotope of carbon, to carbon-12, which is not radioactive, can be used to determine the age of such artifacts. The ratio decreases according to carbon-14's half-life so that the ratio is decreased by one-half every time a half-life passes.

108. 2 The sum of the mass numbers on the left side of a nuclear equation must equal the sum of the mass numbers on the right side of the equation. The same must be true for the sum of the atomic numbers of all species on the left and right side of the equation. On the left, the sum of the mass numbers is 10, and on the right, He has a mass number of 4. The missing species must have a mass number of 6. The sum of the atomic numbers on the left is 5, and He has an atomic number of 2. The missing species must have an atomic number of 3. The symbol for this species is ^6_3Li.

109. 3 Particle accelerators work by using a series of electric and magnetic fields to attract and repel subatomic charged particles. Neutrons carry no charge and are not affected by the fields used in the accelerator.

110. 4 Because the nuclei of atoms contain positively charged protons tightly packed together, when the nuclei of two atoms approach each other

the repulsive forces between the nuclei are extremely large (like charges repel).

111. 4   By using a radioisotope as a tracer, the physical movement of that isotope can easily be traced and information about the mechanism of a reaction, or the uptake of a certain chemical in the body, can be monitored. Choices 1 and 2 describe the process of irradiation and work on the principle that controlled exposure to radioactivity can mutate the DNA in cells causing the cells to die. Choice 3 describes the process of radiological dating, which uses the decay rate of certain isotopes to determine the age of objects.

Group 12—Laboratory Activities

112. 2   Choice 2 (0.125) contains a total of three significant figures, 0.012 has two significant figures, 1,205 has four significant figures, and 12,050 has four significant figures.

113. 1   Take the reading from the bottom of the curved meniscus. Because the bottom of the meniscus lies between the numbers 75 ml and 76 ml, the answer is between those two numbers. The correct answer is choice 1 (and *not* choice 2) because it has the correct number of significant figures. The last significant digit should represent the uncertainty in the reading. In this case, the uncertainty is in the tenths of milliliters, because this is the digit that had to be guessed.

114. 3   Density = mass/volume,
Mass of water = (mass of cylinder + water) − mass of empty cylinder

= 163 g − 141 g

= 22 g, and

Volume of water = 25.6 ml.

Therefore, the density is D = 22 g/25.6 ml = 0.87 g/ml.

115. 3   This problem can be done with only a few steps.

$0.05 \times 36.0\% = 1.8$

(Five percent of the accepted value tells you the deviation of the experimental value.)

$37.8 = 36.0 + 1.8$

Choice 3 accurately shows this relationship between accepted and experimental values.

116. 4 When concentrated sulfuric acid is diluted, it must be mixed with water. The mixing of sulfuric acid and water produces heat. To safeguard against splattering sulfuric acid, it is better to add acid to water instead of adding water to acid. Therefore, choice 4 is the correct answer because it accurately describes the safest technique for diluting concentrated sulfuric acid.

EXAMINATION
JUNE 1993

PART 1: *Answer all 56 questions in this part.* [65]

DIRECTIONS **(1–56):** For each *statement or question, select the word or expression that, of those given, best completes the statement or answers the question. Record your answer on the separate answer sheet provided.*

---

1 If two systems at different temperatures have contact with each other, heat will flow from the system at
 (1) 20.°C to a system at 303 K
 (2) 30.°C to a system at 313 K
 (3) 40.°C to a system at 293 K
 (4) 50.°C to a system at 333 K

2 The graph below represents the uniform heating of a solid, starting below its melting point.

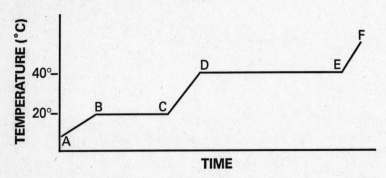

Which portion of the graph shows the solid and liquid phases of the substance existing in equilibrium?
 (1) *AB* (2) *BC* (3) *CD* (4) *DE*

3 What occurs when the temperature of 10.0 grams of water is changed from 15.5°C to 14.5°C?
1 The water absorbs 10.0 calories.
2 The water releases 10.0 calories.
3 The water absorbs 155 calories.
4 The water releases 145 calories.

4 Under the same conditions of temperature and pressure, a liquid differs from a gas because the particles of the liquid
1 are in constant straight-line motion
2 take the shape of the container they occupy
3 have no regular arrangement
4 have stronger forces of attraction between them

5 Compared to the mass of an SO_2 molecule, the mass of an O_2 molecule is
1 one-fourth as great   3 the same
2 one-half as great   4 twice as great

6 Under which conditions does a real gas behave most nearly like an ideal gas?
1 high pressure and low temperature
2 high pressure and high temperature
3 low pressure and low temperature
4 low pressure and high temperature

7 Which statement best describes an electron?
1 It has a smaller mass than a proton and a negative charge.
2 It has a smaller mass than a proton and a positive charge.
3 It has a greater mass than a proton and a negative charge.
4 It has a greater mass than a proton and a positive charge.

8 Which principal energy level has no f sublevel?
   (1) 5   (2) 6   (3) 3   (4) 4

9 What is the mass number of an atom that contains 28 protons, 28 electrons, and 34 neutrons?
   (1) 28   (2) 56   (3) 62   (4) 90

10 In an experiment, alpha particles were used to bombard gold foil. As a result of this experiment, the conclusion was made that the nucleus of an atom is
   1 smaller than the atom and positively charged
   2 smaller than the atom and negatively charged
   3 larger than the atom and positively charged
   4 larger than the atom and negatively charged

11 Given the reaction: $^{131}_{53}I \rightarrow {}^{131}_{54}Xe + X$
   Which particle is represented by X?
   1 alpha   3 neutron
   2 beta   4 proton

12 Which orbital notation represents an atom in the ground state with 6 valence electrons?

13 A white crystalline salt conducts electricity when it is melted and when it is dissolved in water. Which type of bond does this salt contain?

1 ionic 3 covalent
2 metallic 4 network

14 Which diagram best represents the structure of a water molecule?

(1)

(3) O—H—O

(2)

(4) H—H—O

15 What is the total number of moles of oxygen atoms present in 1 mole of $Mg(ClO_3)_2$?

(1) 5 (3) 3
(2) 2 (4) 6

16 Which bond has the greatest ionic character?

(1) H—Cl (3) H—O
(2) H—F (4) H—N

17 Which type of bonding accounts for the unusually high boiling point of water?

1 ionic bonding 3 hydrogen bonding
2 covalent bonding 4 network bonding

18 Which is the correct formula for carbon (II) oxide?
   (1) CO   (3) C_2O
   (2) CO_2 (4) C_2O_3

19 Based on Reference Table G, which of the following
   compounds is *least* stable?
   (1) CO(g)   (3) HF(g)
   (2) CO_2(g)   (4) HI(g)

20 Which electronegativity is possible for an alkali metal?
   (1) 1.0   (2) 2.0   (3) 3.0   (4) 4.0

21 When metals form ions, they tend to do so by
   1 losing electrons and forming positive ions
   2 losing electrons and forming negative ions
   3 gaining electrons and forming positive ions
   4 gaining electrons and forming negative ions

22 Boron and arsenic are similar in that they both
   1 have the same ionization energy
   2 have the same covalent radius
   3 are in the same family of elements
   4 are metalloids (semimetals)

23 Group 18 (0) elements Kr and Xe have selected oxidation states of other than zero. These oxidation states are
   an indication that these elements have
   1 no chemical reactivity
   2 some chemical reactivity
   3 stable nuclei
   4 unstable nuclei

24 The color of Na_2CrO_4 is due to the presence of
   1 a noble gas
   2 a halogen
   3 a transition element
   4 an alkali metal

25 Given the reaction:
   $$Ca + 2H_2O \rightarrow Ca(OH)_2 + H_2$$
   What is the total number of moles of Ca needed to react completely with 4.0 moles of H_2O?
   (1) 1.0   (3) 0.50
   (2) 2.0   (4) 4.0

26 The percent by mass of Ca in $CaCl_2$ is equal to
   (1) $\frac{40}{111} \times 100$ (3) $\frac{3}{1} \times 100$
   (2) $\frac{111}{40} \times 100$ (4) $\frac{1}{3} \times 100$

27 What is the total mass of 3.01×10^{23} atoms of helium gas?
   (1) 8.00 g   (3) 3.50 g
   (2) 2.00 g   (4) 4.00 g

28 Given the reaction:
   $$2C_2H_6(g) + 7O_2(g) \rightarrow 4CO_2(g) + 6H_2O(g)$$
   What is the total number of liters of carbon dioxide formed by the complete combustion of 28.0 liters of $C_2H_6(g)$?
   (1) 14.0 L   (3) 56.0 L
   (2) 28.0 L   (4) 112 L

29 When sodium chloride is dissolved in water, the resulting solution is classified as a
 1 heterogeneous compound
 2 homogeneous compound
 3 heterogeneous mixture
 4 homogeneous mixture

30 According to Reference Table D, a temperature change from 60°C to 90°C has the *least* effect on the solubility of
 (1) SO_2 (3) KCl
 (2) NH_3 (4) $KClO_3$

31 Which series of physical changes represents an entropy increase during each change?
 1 gas → liquid → solid
 2 liquid → gas → solid
 3 solid → gas → liquid
 4 solid → liquid → gas

32 Given the reaction at equilibrium:

$$2H_2(g) + O_2(g) \rightleftarrows 2H_2O(g) + heat$$

Which concentration changes occur when the temperature of the system is increased?
 1 The $[H_2]$ decreases and the $[O_2]$ decreases.
 2 The $[H_2]$ decreases and the $[O_2]$ increases.
 3 The $[H_2]$ increases and the $[O_2]$ decreases.
 4 The $[H_2]$ increases and the $[O_2]$ increases.

33 The change of reactants into products will always be spontaneous if the products, compared to the reactants, have
1 lower enthalpy and lower entropy
2 lower enthalpy and higher entropy
3 higher enthalpy and lower entropy
4 higher enthalpy and higher entropy

Base your answers to questions 34 and 35 on the potential energy diagram of a chemical reaction shown below.

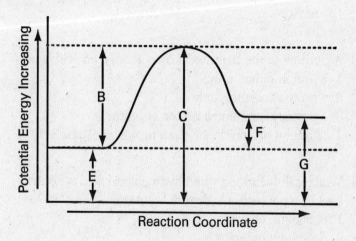

34 Which interval represents the heat of reaction (ΔH)?
(1) E (2) F (3) C (4) G

35 Interval B represents the
1 potential energy of the products
2 potential energy of the reactants
3 activation energy
4 activated complex

36 Based on Reference Table *L*, which solution best conducts electricity?
   (1) 0.1 M HCl   (3) 0.1 M H_2S
   (2) 0.1 M CH_3COOH (4) 0.1 M H_3PO_4

37 Based on Reference Table *E*, a 1-gram quantity of which salt, when placed in 250 milliliters of water and stirred, will produce a solution with the greatest electrical conductivity?
   (1) AgI   (3) $AgNO_3$
   (2) AgCl   (4) Ag_2CO_3

38 According to the Brönsted-Lowry theory, an acid is
   1 a proton donor, only
   2 a proton acceptor, only
   3 a proton donor and a proton acceptor
   4 neither a proton donor nor a proton acceptor

39 Which salt is formed when hydrochloric acid is neutralized by a potassium hydroxide solution?
   1 potassium chloride
   2 potassium chlorate
   3 potassium chlorite
   4 potassium perchlorate

40 Given the equation:

$$H_2O + HF \rightleftarrows H_3O^+ + F^-$$

Which pair represents Brönsted-Lowry acids?
   (1) HF and F^- (3) H_2O and F^-
   (2) HF and H_3O^+ (4) H_2O and H_3O^+

41 What is the pH of a solution that has an OH⁻ ion concentration of 1×10^{-5} mole per liter ($K_w = 1 \times 10^{-14}$)?
  (1) 1   (3) 7
  (2) 5   (4) 9

42 Which half-reaction correctly represents reduction?
  (1) $Fe^{2+} + 2e^- \rightarrow Fe$
  (2) $Fe^{2+} + e^- \rightarrow Fe^{3+}$
  (3) $Fe + 2e^- \rightarrow Fe^{2+}$
  (4) $Fe + e^- \rightarrow Fe^{3+}$

43 In which compound does hydrogen have an oxidation number of –1?
  (1) NH_3 (3) HCl
  (2) KH   (4) H_2O

44 In the reaction $2H_2(g) + O_2(g) \rightarrow 2H_2O(g)$, the oxidizing agent is
  (1) H_2 (3) H^+
  (2) O_2 (4) O^{2-}

45 Which reaction occurs when a strip of magnesium metal is placed in a solution of $CuCl_2$?
  1 The chloride ion is oxidized.
  2 The chloride ion is reduced.
  3 The magnesium metal is oxidized.
  4 The magnesium metal is reduced.

46 Given the reaction:

$$Zn(s) + Cu^{2+}(aq) \rightarrow Zn^{2+}(aq) + Cu(s)$$

Which particles must be transferred form one reactant to the other reactant?

1 ions   3 protons
2 neutrons   4 electrons

47 Which redox equation is correctly balanced?

(1) $Cr^{3+} + Mg \rightarrow Cr + Mg^{2+}$
(2) $Al^{3+} + K \rightarrow Al + K^+$
(3) $Sn^{4+} + H_2 \rightarrow Sn + 2H^+$
(4) $Br_2 + Hg \rightarrow Hg^{2+} + 2Br^-$

48 Organic chemistry is the chemistry of compounds containing the element

1 carbon   3 nitrogen
2 hydrogen   4 oxygen

49 The isomers CH_3OCH_3 and CH_3CH_2OH differ in

1 molecular formula
2 molecular structure
3 number of atoms
4 formula mass

50 Given the molecule:

Replacing a hydrogen atom on this molecule with the functional group —OH will change the original properties of the molecule to those of an

1 ester   3 acid
2 ether   4 alcohol

51 Which structural formula represents a member of the series of hydrocarbons having the general formula C_nH_{2n-2}?

(1)
$$H-\overset{\overset{\displaystyle H}{|}}{\underset{\underset{\displaystyle H}{|}}{C}}-\overset{\overset{\displaystyle H}{|}}{\underset{\underset{\displaystyle H}{|}}{C}}-H$$

(2)
$$\overset{H}{\underset{H}{>}}C=C\overset{H}{\underset{H}{<}}$$

(3) $H-C\equiv C-H$

(4)
$$H-\overset{\overset{\displaystyle H}{|}}{\underset{\underset{\displaystyle H}{|}}{C}}-\overset{\overset{\displaystyle H}{|}}{C}=C\overset{H}{\underset{H}{<}}$$

52 What is the total number of valence electrons in a carbon atom in the ground state?
(1) 6   (3) 12
(2) 2   (4) 4

Note that questions 53 through 56 have only three choices.

53 As the elements of Group 17 (VIIA) are considered in order of increasing atomic number, the nonmetallic character of each successive element
1 decreases
2 increases
3 remains the same

54 As the atoms of the metals of Group 1 (IA) in the ground state are considered in order from top to bottom, the number of occupied principal energy levels
1 decreases
2 increases
3 remains the same

55 As the mass number of the isotopes of hydrogen increases, the number of protons
1 decreases
2 increases
3 remains the same

56 As $Cu(NO_3)_2$ is dissolved in pure water, the pH of the resulting solution
1 decreases
2 increases
3 remains the same

PART 2: *This part consists of twelve groups. Choose seven of these twelve groups. Be sure to answer all questions in each group chosen. Write the answers to these questions on the separate answer sheet provided.* [35]

GROUP 1—Matter and Energy

If you choose this group, be sure to answer questions 57–61.

57 Which phase change represents sublimation?
1 solid → gas   3 gas → solid
2 solid → liquid   4 gas → liquid

58 Which property of a sample of mercury is different at 320 K than at 300 K?

1 atomic mass
2 atomic radius
3 vapor pressure
4 melting point

59 Which statement describes a chemical property of the element iodine?

1 Its crystals are a metallic gray.
2 It dissolves in alcohol.
3 It forms a violet-colored gas.
4 It reacts with hydrogen to form a gas.

60 The characteristic which distinguishes a true solid from other phases of matter at STP is that in a true solid, the particles are

1 vibrating and changing their relative positions
2 vibrating without changing their relative positions
3 motionless but changing their relative positions
4 motionless without changing their relative positions

61 The volume of a given mass of an ideal gas at constant pressure is

1 directly proportional to the Kelvin temperature
2 directly proportional to the Celsius temperature
3 inversely proportional to the Kelvin temperature
4 inversely proportional to the Celsius temperature

GROUP 2—Atomic Structure

If you choose this group, be sure to answer questions 62–66.

62 Neutral atoms of ^{35}Cl and ^{37}Cl differ with respect to their number of

1 electrons   3 neutrons

2 protons   4 positrons

63 What is the total number of electrons present in an atom of $^{59}_{27}Co$?

(1) 27   (3) 59

(2) 32   (4) 86

64 Which of the following atoms has the greatest nuclear charge?

(1) Al   (3) Si

(2) Ar   (4) Na

65 An element has an atomic number of 18. What is the principal quantum number (n) of its outermost electrons?

(1) 1   (3) 3

(2) 2   (4) 4

66 What is the total mass of ^{222}Rn remaining in an original 160-milligram sample of ^{222}Rn after 19.1 days?

(1) 2.5 mg   (3) 10. mg

(2) 5.0 mg   (4) 20. mg

GROUP 3—Bonding

*If you choose this group, be sure to answer questions **67–71**.*

67 In a nonpolar covalent bond, electrons are
   1 located in a mobile "sea" shared by many ions
   2 transferred from one atom to another
   3 shared equally by two atoms
   4 shared unequally by two atoms

68 Which compound has the same empirical and molecular formula?
   1 acetylene   3 ethane
   2 ethene   4 methane

69 When the equation
   $\_C_8H_{16} + \_O_2 \rightarrow \_CO_2 + \_H_2O$ is correctly balanced using the smallest whole number coefficients, the coefficient of O_2 is
   (1) 1   (3) 12
   (2) 8   (4) 16

70 Which species can form a coordinate covalent bond with an H^+ ion?
   (1) H·   (3) H^+
   (2) H:⁻   (4) H : H

71 In which chemical system are molecule-ion attractions present?
   (1) KCl(g)   (3) KCl(s)
   (2) KCl(ℓ)   (4) KCl(aq)

GROUP 4—Periodic Table

If you choose this group, be sure to answer questions **72–76.**

72 Which atom has a radius larger than the radius of its ion?

(1) Cl
(2) Ca
(3) S
(4) Se

73 The chemical properties of the elements are periodic functions of their atomic

1 masses
2 weights
3 numbers
4 radii

74 Which of the following substances is the best conductor of electricity?

(1) NaCl(s)
(2) Cu(s)
(3) $H_2O(\ell)$
(4) $Br_2(\ell)$

75 Which halogen is a solid at STP?

1 fluorine
2 chlorine
3 bromine
4 iodine

76 Atoms of nonmetals generally react with atoms of metals by

1 gaining electrons to form ionic compounds
2 gaining electrons to form covalent compounds
3 sharing electrons to form ionic compounds
4 sharing electrons to form covalent compounds

GROUP 5—Mathematics of Chemistry

If you choose this group, be sure to answer questions **77–81.**

77 What is the empirical formula of a compound composed of 2.8% by mass of boron and 97% by mass of iodine?
- (1) BI_2
- (2) B_2I
- (3) BI_3
- (4) B_3I

78 A gas has a volume of 2 liters at 323 K and 3 atmospheres. When its temperature is changed to 273 K and the pressure is changed to 1 atmosphere, the new volume of the gas would be equal to

- (1) $2\,L \times \dfrac{273\text{ K}}{323\text{ K}} \times \dfrac{1\text{ atm}}{3\text{ atm}}$

- (2) $2\,L \times \dfrac{323\text{ K}}{273\text{ K}} \times \dfrac{1\text{ atm}}{3\text{ atm}}$

- (3) $2\,L \times \dfrac{273\text{ K}}{323\text{ K}} \times \dfrac{3\text{ atm}}{1\text{ atm}}$

- (4) $2\,L \times \dfrac{323\text{ K}}{273\text{ K}} \times \dfrac{3\text{ atm}}{1\text{ atm}}$

79 Which gas could have a density of 2.05 grams per liter at STP?
- (1) N_2O_3
- (2) NO_2
- (3) HF
- (4) HBr

80 What is the total number of calories of heat energy absorbed when 10.0 grams of water is vaporized at its normal boiling point?
- (1) 7.97
- (2) 53.9
- (3) 5390
- (4) 7970

81 How many moles of a nonvolatile, nonelectrolyte solute
   are required to lower the freezing point of 1,000 grams
   of water by 5.58°C?
   (1) 1   (3) 3
   (2) 2   (4) 4

GROUP 6—Kinetics and Equilibrium

If you choose this group, be sure to answer questions 82–86.

82 Based on Reference Table *G*, which compound forms spontaneously even though the ΔH for its formation is positive?
   (1) $C_2H_4(g)$ (3) $ICl(g)$
   (2) $C_2H_2(g)$ (4) $HI(g)$

83 Given the reaction at equilibrium:

   $$AgCl(s) \rightleftarrows Ag^+(aq) + Cl^-(aq)$$

   The addition of Cl^- ions will cause the concentration of
   $Ag^+(aq)$ to
   1 decrease as the amount of $AgCl(s)$ decreases
   2 decrease as the amount of $AgCl(s)$ increases
   3 increase as the amount of $AgCl(s)$ decreases
   4 increase as the amount of $AgCl(s)$ increases

84 The expression $\Delta H - T\Delta S$ is equal to the change in
   1 binding energy   3 free energy
   2 ionization energy   4 activation energy

85 At room temperature, which reaction would be
   expected to have the fastest reaction rate?
   (1) $Pb^{2+}(aq) + S^{2-}(aq) \rightarrow PbS(s)$
   (2) $2H_2(g) + O_2(g) \rightarrow 2H_2O(\ell)$
   (3) $N_2(g) + 2O_2(g) \rightarrow 2NO_2(g)$
   (4) $2KClO_3(s) \rightarrow 2KCl(s) + 3O_2(g)$

86 Which statement is true for a saturated solution?
   1 It must be a concentrated solution.
   2 It must be a dilute solution.
   3 Neither dissolving nor crystallizing is occurring.
   4 The rate of dissolving equals the rate of crystallizing.

GROUP 7—Acids and Bases

If you choose this group, be sure to answer questions 87–91.

87 When an Arrhenius base is placed in H_2O, the only negative ion present in the solution is
   (1) OH^-
   (2) H_3O^-
   (3) H^-
   (4) O^{2-}

88 Which solution will change red litmus to blue?
   (1) $HCl(aq)$
   (2) $NaCl(aq)$
   (3) $CH_3OH(aq)$
   (4) $NaOH(aq)$

89 A chloride ion, $[:\ddot{C}l:]^-$ acts as a Brönsted base when it combines with
   1 an OH^- ion
   2 a K^+ ion
   3 an H^- ion
   4 an H^+ ion

90 Which equation illustrates amphoterism?
   (1) $NaCl \rightarrow Na^+ + Cl^-$
   (2) $NaOH \rightarrow Na^+ + OH^-$
   (3) $H_2O + H_2O \rightarrow H_3O^+ + OH^-$
   (4) $HCl + H_2O \rightarrow H_3O^+ + Cl^-$

91 In a titration, the endpoint of a neutralization reaction was reached when 37.6 milliliters of an HCl solution was added to 17.3 milliliters of a 0.250 M NaOH solution. What was the molarity of the HCl solution?

(1) 0.115 M   (3) 0.250 M

(2) 0.203 M   (4) 0.543 M

GROUP 8—Redox and Electrochemistry

If you choose this group, be sure to answer questions 92–96.

Base your answers to questions 92 and 93 on the diagram below which represents an electrochemical cell.

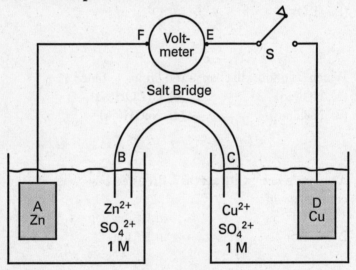

92 Which statement correctly describes the direction of flow for the ions in this cell when the switch is closed?

1 Ions move through the salt bridge from *B* to *C*, only.

2 Ions move through the salt bridge from *C* to *B*, only.

3 Ions move through the salt bridge in both directions.
4 Ions do not move through the salt bridge in either direction.

93 When the switch is closed, which group of letters correctly represents the direction of electron flow?
(1) $A \rightarrow B \rightarrow C \rightarrow D$
(3) $D \rightarrow C \rightarrow B \rightarrow A$
(2) $A \rightarrow F \rightarrow E \rightarrow D$
(4) $D \rightarrow E \rightarrow F \rightarrow A$

94 Based on Reference Table N, which metal will react with H^+ ions to produce $H_2(g)$?
(1) Au
(3) Cu
(2) Ag
(4) Mg

95 What is the standard electrode (E^0) assigned to the half-reaction $Cu^{2+} + 2e^- \rightarrow Cu(s)$ when compared to the standard hydrogen half-reaction?
(1) +0.34 V
(3) +0.52 V
(2) –0.34 V
(4) –0.52 V

96 Which species acts as the anode when the reaction $Zn(s) + Pb^{2+}(aq) \rightarrow Zn^{2+}(aq) + Pb(s)$ occurs in an electrochemical cell?
(1) $Zn(s)$
(3) $Pb^{2+}(aq)$
(2) $Zn^{2+}(aq)$
(4) $Pb(s)$

GROUP 9—Organic Chemistry

If you choose this group, be sure to answer questions **97–101**.

97 The products of condensation polymerization are a polymer and
1 carbon dioxide
2 water
3 ethanol
4 glycerol

98 Given the equation:

$$C_6H_{12}O_6 \rightarrow 2C_2H_5OH + 2CO_2$$

The chemical process illustrated by this equation is
1 fermentation
2 saponification
3 esterification
4 polymerization

99 Which two compounds are monohydroxy alcohols?
1 ethylene glycol and ethanol
2 ethylene glycol and glycerol
3 methanol and ethanol
4 methanol and glycerol

100 Which type of compound is represented by the structural formula shown below?

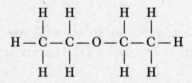

1 a ketone
2 an aldehyde
3 an ester
4 an ether

101 Which is the structural formula for 2-chlorobutane?

(1)
```
      H   H   H   Cl
      |   |   |   |
  H — C — C — C — C — H
      |   |   |   |
      H   H   H   Cl
```

(2)
```
      H   H   H   Cl
      |   |   |   |
  H — C — C — C — C — H
      |   |   |   |
      H   H   H   H
```

(3)
```
      H   Cl  H   H
      |   |   |   |
  H — C — C — C — C — H
      |   |   |   |
      H   H   H   H
```

(4)
```
      H   H   Cl  H
      |   |   |   |
  H — C — C — C — C — H
      |   |   |   |
      H   H   Cl  H
```

GROUP 10—**Applications of Chemical Principles**

*If you choose this group, be sure to answer questions **102–106**.*

102 Given the lead-acid battery reaction:

$$Pb + PbO_2 + 2H_2SO_4 \underset{\text{charge}}{\overset{\text{discharge}}{\rightleftharpoons}} 2PbSO_4 + 2H_2O$$

When the reaction produces electricity, which element changes oxidation states?
(1) Pb   (3) H
(2) O   (4) S

103 During fractional distillation of petroleum, which of the following fractions has the *lowest* boiling point?
(1) C_8H_{18} (3) $C_{15}H_{32}$
(2) $C_{12}H_{26}$ (4) $C_{18}H_{38}$

104 What is the original source of many textiles and most plastics?
1 coal   3 petroleum
2 wood   4 mineral ores

105 Given a reaction that occurs in the contact process:

$$2SO_2(g) + O_2(g) \rightleftarrows 2SO_3(g) + heat$$

Adding a catalyst to this system causes the
1 activation energy to decrease
2 activation energy to increase
3 heat of reaction to decrease
4 heat of reaction to increase

106 Given the reaction:

$$2Al + Cr_2O_3 \rightarrow Al_2O_3 + 2Cr$$

When this reaction is used to produce chromium, the aluminum is acting as
1 a catalyst   3 an oxidizing agent
2 an alloy   4 a reducing agent

GROUP 11—Nuclear Chemistry

If you choose this group, be sure to answer questions 107–111.

107 The diagram below shows a nuclear reaction in which a neutron is captured by a heavy nucleus.

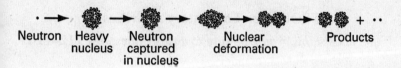

Neutron   Heavy   Neutron   Nuclear   Products
          nucleus captured deformation
                   in nucleus

Which type of reaction is illustrated by the diagram?
1 an endothermic fission reaction
2 an exothermic fission reaction
3 an endothermic fusion reaction
4 an exothermic fusion reaction

108 Heavy water and graphite are two examples of materials that can be used in a nuclear reactor to slow down neutrons. These materials are called
1 fuels   3 coolants
2 shields   4 moderators

109 Which particle can *not* be accelerated in a magnetic field?
1 alpha particle   3 neutron
2 beta particle   4 proton

110 Which radioisotope is used to diagnose thyroid disorders?
1 cobalt-60   3 technetium-99
2 iodine-131   4 uranium-238

111 Given the transmutation:

$$_{0}^{1}n + _{92}^{235}U \rightarrow _{56}^{141}Ba + X + 3_{0}^{1}n$$

The element X has an atomic number of
(1) 36
(3) 92
(2) 89
(4) 93

GROUP 12—Laboratory Activities
If you choose this group, be sure to answer questions 112–116.

112 Which diagram represents a test tube holder (clamp)?

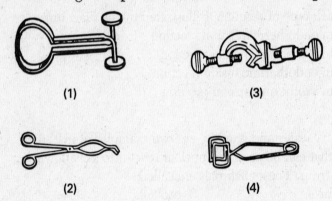

(1)

(3)

(2)

(4)

113 In a laboratory exercise to determine the volume of a mole of a gas at STP, a student determines the volume to be 2.25 liters greater than the accepted value of 22.4 liters. The percent error in the student's value is
(1) 2.25%
(3) 20.2%
(2) 10.0%
(4) 24.7%

114 To determine the density of an irregularly shaped object, a student immersed the object in 21.2 milliliters of H_2O in a graduated cylinder, causing the level of the H_2O to ride to 27.8 milliliters. If the object had a mass of 22.4 grams, what was the density of the object?

(1) 27.8 g/mL   (3) 3.0 g/mL

(2) 6.6 g/mL   (4) 3.4 g/mL

115 A material will be used to fill an empty beaker to level A, as shown in the diagram below.

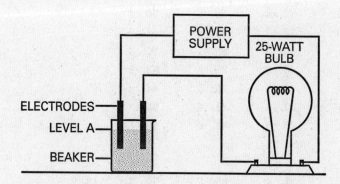

Which material, when used to fill the beaker to level A, would cause the bulb to glow brightly?

(1) $C_6H_{12}O_6(s)$ (3) KCl(s)

(2) $C_6H_{12}O_6(aq)$ (4) KCl(aq)

116 A solid is dissolved in a beaker of water. Which observation suggests that the process is endothermic?

1 The solution gives off a gas.

2 The solution changes color.

3 The temperature of the solution decreases.

4 The temperature of the solution increases.

ANSWER KEY
JUNE 1993

PART 1

1. 3	15. 4	29. 4	43. 2
2. 2	16. 2	30. 1	44. 2
3. 2	17. 3	31. 4	45. 3
4. 4	18. 1	32. 4	46. 4
5. 2	19. 4	33. 2	47. 4
6. 4	20. 1	34. 2	48. 1
7. 1	21. 1	35. 3	49. 2
8. 3	22. 4	36. 1	50. 4
9. 3	23. 2	37. 3	51. 3
10. 1	24. 3	38. 1	52. 4
11. 2	25. 2	39. 1	53. 1
12. 2	26. 1	40. 2	54. 2
13. 1	27. 2	41. 4	55. 3
14. 1	28. 3	42. 1	56. 1

PART 2

57. 1	71. 4	85. 1	99. 3	113. 2
58. 3	72. 2	86. 4	100. 4	114. 4
59. 4	73. 3	87. 1	101. 3	115. 4
60. 2	74. 2	88. 4	102. 1	116. 3
61. 1	75. 4	89. 4	103. 1	
62. 3	76. 1	90. 3	104. 3	
63. 1	77. 3	91. 1	105. 1	
64. 2	78. 3	92. 3	106. 4	
65. 3	79. 2	93. 2	107. 2	
66. 2	80. 3	94. 4	108. 4	
67. 3	81. 3	95. 1	109. 3	
68. 4	82. 3	96. 1	110. 2	
69. 3	83. 2	97. 2	111. 1	

ANSWERS AND EXPLANATIONS
JUNE 1993

PART 1

1. **3** Heat always flows passively from objects that are hotter to objects that are cooler. Because the answer choices contain temperatures in both Celsius and Kelvin, first convert all temperatures in Celsius to Kelvin using the formula: K = °C + 273. Once the conversion has been made, it is easier to compare the given temperatures.

Answer choice	First temperature (K)	Second temperature (K)
1	20 + 273 = 292	303
2	30 + 273 = 303	313
3	40 + 273 = 313	293
4	50 + 272 = 323	333

2. **2** The portion of the graph labeled *BC* represents the area where solid and liquid phases of the substances exist in equilibrium. Solid and liquid phases only exist in equilibrium when melting takes place. Look at the phase diagram below for an explanation of phase changes.

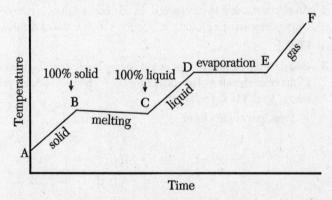

3. **2** When the water temperature is lowered, energy is released by the water molecules. In this problem, water temperature was lowered from 15.5°C to 14.5°C. Therefore, energy must have been released

when the water cooled to a lower temperature. The amount of energy (in calories) released from water can be calculated by using the following formula:

Energy or heat = mass of water × change in temperature °C × specific heat of water

$$= 10 \text{ g} \times 1°C \times 1 \text{ calorie/g°C}$$
$$= 10 \text{ calories}$$

4. 4   According to the kinetic theory of gases, there are no attractive forces between gas molecules. On the other hand, attractive forces, such as hydrogen bonding, are present between liquid molecules. Therefore, liquids differ from gases because attractive forces exist between liquid particles. All the other characteristics mentioned in choices 1, 2, and 3 are true for both gases and liquids.

5. 2   The mass of SO_2 is 32 (sulfur) + 32 (oxygen [O_2] = 16 × 2) = 64 g. The mass of O_2 (oxygen) is 32 g. Therefore, the mass of an O_2 molecule is one-half as great as the mass of an SO_2 molecule.

6. 4   Gases behave ideally when pressure is low and the temperature is high. When the pressure is low and the temperature is high, no attractive forces exist between gas molecules. According to the kinetic theory of gases, ideal gases lack any attractive forces between ideal gas particles.

7. 1   An electron is negatively charged. In addition, its mass is minute relative to the mass of a proton. An electron's mass is $\frac{1}{1836}$ of the mass of a proton.

8. 3   Based on the table below, principal energy levels 4 and higher contain f sublevels. Therefore, the correct answer is choice 3 because principal energy level 3 lacks f sublevels.

Principal energy level	Sublevels
1	s
2	s, p
3	s, p, d
4	s, p, d, f

9. 3   The mass number is the sum of all the protons plus all the neutrons in an atom's nucleus.

Mass number = protons + neutrons
$$= 28 + 34$$
$$= 62$$

10. 1 The experiment described in the question is better known as the *Rutherford experiment*. When Rutherford bombarded a thin piece of gold foil with positively charged alpha particles, most of the particles passed straight through the foil. However, some were deflected, and a few reflected off the foil. From his observations, Rutherford concluded that the atom is mostly empty space, but he also stated that in the center of each atom lies a small nucleus containing positive charges.

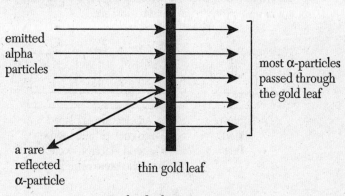

Rutherford's Experiment

11. 2 In the reaction

$$^{131}_{53}\text{I} \rightarrow ^{131}_{54}\text{Xe} + \text{X},$$

there is a change in the atomic number when I (53) changes to Xe (54). The atomic number only changes by +1 (53 to 54) if a beta particle is released. Therefore, choice 2 must be the correct answer. If the atomic number changed by –1 (e.g., 53 to 52), then the released particle would be an alpha particle (choice 1).

$$\text{I} \longrightarrow \text{Xe} + \text{X}$$

I	Xe	X
mass #131	mass #131	mass #0
charge 53	charge 54	charge –1

Overall, masses and charges must be balanced on each side of the equation

12. **2** Valence electrons are found in the outermost energy level. Valence electrons are found in either the *s* or *p* subshell of the outermost or valence shell (energy level). Because there are six valence electrons, two electrons must be present in the *s* subshell and four electrons must be present in the *p* subshell. Recall that the *s* subshell can hold up to two electrons and the *p* subshell can hold up to six electrons. However, the *s* subshell must be completely filled before electrons can be placed in the *p* subshell. We also know that an electron must be placed in each one of the three orbitals of the *p* subshell before a second electron can be placed in any one of the three orbitals. Based on this fact, only choice 2 correctly depicts the placement of six valence electrons in the *s* and *p* subshells.

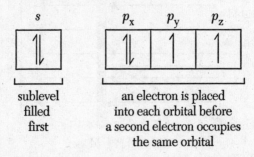

13. **1** To conduct electricity when dissolved in water, a compound must be an electrolyte. Electrolytes dissociate into charged particles or ions when dissolved in water. Any compound that produces ions in water is called an *ionic compound*. Ionic compounds also conduct electricity when molten because the ions are free to move. All ionic compounds contain ionic bonds. Salt is an example of an ionic compound. Ionic compounds are composed of a metal and a nonmetal. Based on the characteristics of ionic compounds, we can conclude that the correct answer is choice 1.

14. **1** Water's molecular formula is H_2O. This means that a water molecule is composed of two hydrogen atoms and one oxygen atom. Because hydrogen can only form one bond and oxygen usually forms two covalent bonds, the only correct structure is choice 1, which also correctly indicates the bent shape of the water molecule due to the two lone pairs of electrons on the oxygen atom.

15. **4** Count the number of oxygen atoms present in a molecule of $Mg(ClO_3)_2$. This is the number of moles of oxygen present in a mole of $Mg(ClO_3)_2$. Therefore, the number of moles of oxygen atoms is

Oxygen atoms = $Mg(ClO_3)_2 = 3 \times 2 = 6$

16. **2** The bond with the highest electronegativity difference has the greatest ionic character. To compute the electronegativity differences of atoms, Reference Table K must be used.

Choice	Bond	Electronegativity difference
1	H—Cl	1.0
2	H—F	1.8
3	H—O	1.2
4	H—N	0.9

Based on the data, choice 2 is the correct answer because H and F have the greatest electronegativity difference.

17. **3** Presence of weak intermolecular forces between water molecules is responsible for water's unusually high boiling point. Of all the choices given, only hydrogen bonding is an example of a weak intermolecular force. Choices 1, 2, and 4 are examples of *intra*molecular forces of attraction.

water molecule

hydrogen bond

18. **1** In the formula for carbon (II) oxide, carbon (II) represents a carbon atom with an oxidation number of +2. The oxidation number of oxy-

gen is –2. Because positive and negative charges (in this question +2 and –2 charges) cancel each other out in a neutral molecule, the correct formula of carbon (II) oxide must be CO (choice 1).

$$C^{+2} \diagdown \diagup O^{-2}$$

cross-multiply by 2 to balance the charges

$$C_2O_2 = CO \quad \text{[empirical formula]}$$

19. 4 Reference Table G lists the heat of formation (ΔH°_f) of various compounds. The less energy a compound has, the more stable it is. In other words, a stable compound is formed if energy is released during the formation of that compound. If energy is released during formation, ΔH°_f is negative. On the other hand, an unstable product compound is formed if energy is absorbed during the formation of that compound. When energy is absorbed during formation, ΔH°_f is positive. Therefore, from Table G pick the compound that has a positive ΔH°_f. Of the four answer choices given, only choice 4 (HI) has a positive ΔH°_f of +6.3 kcal/mole.

20. 1 Alkali metals belong to group 1 of the periodic table. All alkali metals ionize to form a positively charged ion and have low electronegativity values. According to Reference Table K, alkali metals have electronegative values between 0.7 and 1.0. Lithium, an alkali metal, has an electronegative value of 1.0.

21. 1 Metals have very low ionization energies as well as electronegativities. As a result, it is easy for a metals to lose electrons from its valence shell. When a metal forms an ion, it loses electrons from its valence shell. Consequently, the ion that is formed has more protons than electrons in it. Because there is a greater number of positive than negative charges in ions formed by metals, they have positive charges. The answer choice that correctly states that positively charged metal ions are formed as a result of the loss of electrons is answer choice 1.

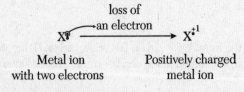

22. **4** Boron and arsenic have different ionization energies (see Reference Table K) and covalent radii (see Reference Table P). In addition, boron belongs to group 13 and arsenic belongs to group 15 of the periodic table. Based on this information, choices 1, 2, and 3 are not the correct answers. However, both boron and arsenic are metalloids; they lie next to the bold line that divides the metals from nonmetals in the periodic table.

23. **2** Elements can either gain or lose electrons, but they only gain or lose electrons if they participate in chemical reactions such as redox reactions. Group 18 elements, also known as *noble gases,* are chemically inactive and neither lose nor gain electrons. However, the fact that Kr and Xe, members of group 18 of the periodic table, have selective oxidation is an indication that they have some chemical reactivity.

24. **3** Colored salts and colored solutions of salts usually indicate the presence of a transition element. Transition elements are found in groups 2–11 of the periodic table. In the compound Na_2CrO_4, chromium (Cr) is the transition element. The d electrons in transition elements are responsible for their colors.

25. **2** According to the balanced reaction $Ca + 2H_2O \rightarrow Ca(OH)_2 + H_2$, for every 1 mole of Ca that reacts, two moles of H_2O are consumed. In other words, Ca and H_2O react with each other in a 1:2 ratio. Therefore, if 4 moles of H_2O were to react completely, a total of 2 moles of Ca must be present.

26. **1** Use the following formula to calculate percent by mass of Ca in $CaCl_2$:

% mass = mass of element in compound / mass of compound × 100%
Mass of compound $CaCl_2$: Ca = 1 × 40 = 40
$$Cl = 2 \times 35.5 = 71$$
Total mass of $CaCl_2$ = 111
% Ca by mass = 40 / 111 × 100%

27. **2** One mole of any gas contains 6.02×10^{23} particles. Therefore, 3.01×10^{23} particles (half the number 6.02×10^{23}) are contained in 0.5 moles of helium gas. The atomic mass of a mole of helium is 4 grams, so 0.5 moles of helium has a mass of 4 grams per mole × 0.5 mole = 2 grams.

28. 3 According to the balanced reaction

$$2C_2H_6(g) + 7O_2(g) \rightarrow 4CO_2(g) + 6H_2O(g),$$

mole ratio is 2, 7 → 4, 6, and volume is 28 L → x. Use the mole ratio to solve for the volume of carbon dioxide produced:

$$\frac{2}{28} = \frac{4}{x}$$

$x = 56.0$ L of CO_2 formed.

29. 4 When sodium chloride (NaCl) is dissolved in water, it dissociates into Na^+ and Cl^- ions. The resulting solution is not a new compound; it is a mixture of H_2O and Na^+ and Cl^- ions. The ions dissolve evenly throughout the water. In other words, the concentration of Na^+ and Cl^- ions are the same everywhere in water. Therefore, it is a homogenous solution because all parts of the solution are the same, or uniform.

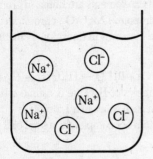 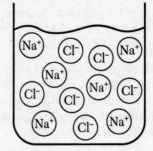

diluted NaCl solution   concentrated NaCl solution

30. 1 Use Reference Table D to answer this question. According to the table, as temperature changes from 60°C to 90°C, the solubility of SO_2 remains the same. On the graph below, the solubility curve is flat for SO_2 between the temperatures 60°C and 90°C. For all other compounds, the solubility curve changes. Therefore, the solubility of SO_2 (choice 1) is least affected by changing temperature.

31. 4 Entropy, or randomness, increases as matter changes from solid to liquid to gas. As entropy increases, molecules gain more energy and, thus, are able to move around more randomly. Choice 4 correctly states the increase in entropy during phase changes.

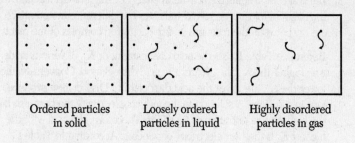

| Ordered particles in solid | Loosely ordered particles in liquid | Highly disordered particles in gas |

32. **4** In the reaction

$$2H_2(g) + O_2(g) \rightarrow 2H_2O(g) + heat,$$

heat is released or acts as a product. In other words, this is an exothermic reaction. Therefore, if temperature increases, the reaction equilibrium shifts to the left because exothermic reactions do not favor high temperatures. As a result, the amounts of H_2 and O_2 increase.

33. **2** When a spontaneous reaction takes place, the reaction is usually exothermic—energy is released in the form of heat. As a result, the products of the reaction have lower potential energy than the reactants. Change in potential energy is also called *change in enthalpy*. On the other hand, an increase in entropy, or randomness, favors spontaneous reactions. Therefore, the two factors that favor a spontaneous reaction are (1) an increase in entropy, and (2) a decrease in enthalpy.

34. **2** The heat of reaction (ΔH) is represented by F. (See the labeled diagram below.) F indicates the difference in potential energy between reactants and products.

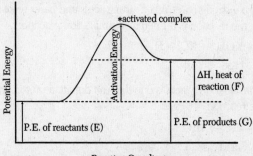

35. 3 Interval B represents the activation energy of the forward reaction. See the diagram for question 34. B indicates the difference in potential energy between the reactants and the activated complex of the reaction.

36. 1 Reference Table L lists the acid dissociation, or K_a, of various acids. The higher the K_a, the greater the acid dissociation. Conversely, the lower the K_a, the lower the acid dissociation. Dissociation produces ions in solutions. A solution that conducts electricity the best also has the highest concentration of ions in it. Therefore, the acid with the highest K_a is the best electricity conductor. According to Table L, HCl has the highest K_a.

37. 3 This question is similar to the last question. In this question, instead of looking at K_a values, solubility constants are used to determine the dissociation rate of salts. Reference Table E lists the solubility of various compounds in water. The higher the solubility, the greater the salt dissociation. Conversely, the lower the solubility, the lower the salt dissociation. Dissociation produces ions in solutions. A solution that conducts electricity the best also has the highest concentration of ions in it. Therefore, the salt with the greatest solubility is the best electricity conductor. According to Table E, $AgNO_3$ has the highest solubility. All other salts are nearly insoluble in water. Therefore, the best conductor of electricity is $AgNO_3$ (choice 3).

38. 1 The Brönsted-Lowry theory states that a substance that can release or donate a proton (H^+) during a reaction is an acid.

$$HX \rightarrow H^+ + X^-$$

39. 1 When hydrochloric acid (HCl) and potassium hydroxide (KOH) react with each other, they form salt and water. HCl is a strong acid and KOH is a strong base. When strong acids and bases react with each other, the reaction is termed a neutralization reaction.

$$HCl + KOH \rightarrow KCl + H_2O$$
acid   base   salt

40. 2 A Brönsted-Lowry acid is one that can donate a proton (H^+).
In the reaction

$$H_2O + HF \rightarrow H_3O^+ + F^-,$$
base   acid   acid   base

the Brönsted-Lowry acid pair is HF and H_3O^+. In the forward reaction, HF donates a proton to water. In the reverse reaction, H_3O^+ donates a proton to F^-.

41. **4** To calculate pH, we need to know the concentration of H^+ in solution. The $[H^+]$ concentration can be determined by using the following relationship:

$K_w = [H^+][OH^-]$

$1 \times 10^{-14} = [1 \times 10^{-5}][H^+]$

$1 \times 10^{-9} = [H^+]$

$pH = -\log [H^+]$

$pH = -\log [1 \times 10^{-9}] = -(-9) = 9$

42. **1** During reduction, electrons are gained by an element or ion. As a result, the oxidation number of an element is always lowered as a result of reduction. Choice 1 is the correct answer because it shows an element gaining electrons and a subsequent reduction in its oxidation number.

$Fe^{+2} + 2e^- \rightarrow Fe^0$

In all other answer choices, the oxidation number increased.

43. **2** Hydrogen's normal oxidation number is +1. In KH, however, hydrogen's oxidation number is –1. Because K (potassium) is a metal and takes a charge of +1, hydrogen must have a charge of –1. In choices 1, 3, and 4, hydrogen's oxidation number is +1.

44. **2** It is important to realize that the oxidizing agent itself undergoes reduction. Therefore, in the given reaction $2H_2 + O_2 \rightarrow 2H_2O$, the oxidizing agent is the species whose oxidation number is lowered following the completion of the reaction. Both hydrogen and oxygen have an oxidation number of zero at the start of the reaction because they are in their elemental form. In water, however, hydrogen has a charge of +1 and oxygen has a charge of –2. Therefore, oxygen undergoes reduction and must be the oxidizing agent.

45. **3** When magnesium (Mg) metal is placed in a $CuCl_2$ solution, the Mg undergoes a single replacement reaction with $CuCl_2$ to form $MgCl_2$. In the process, the elemental Mg with a charge of zero is ionized to

Mg^{+2}. Therefore, Mg must have undergone oxidation because the oxidation number increased.

$Mg + CuCl_2 \rightarrow MgCl_2 + Cu$

Half reaction: $Mg \rightarrow Mg^{+2} + 2e^-$

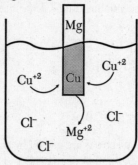

Cu^{+2} (copper) settles on
the metal bar as Mg^{+2} comes off

46. 4   The stated reaction is an example of a reduction-oxidation (redox) reaction. During a redox reaction, electrons are transferred from one reactant to another. Remember that electrons lie outside the nucleus of an atom. Anything located inside the nucleus is incapable of leaving the nucleus, so only electrons can migrate between reactants.

47. 4   A redox reaction is correctly balanced when the total number of atoms of each element and the total charges on the two sides are equal. Choice 4 is the correct answer because the total charge produced on the right is the same as the total charge on the left (zero).

Choice 4: $Br_2 + Hg \rightarrow Hg^{+2} + 2Br^-$
Charges: 0   0   +2 $-2 \ (2 \times -1)$

In all other choices, the total charges are unbalanced on the left and right side of the reactions.

48. 1   All organic compounds contain the element carbon. Therefore, organic chemistry is the study of compounds containing the element carbon.

49. 2   The definition of a pair of isomers is that both the compounds of the pair have the same molecular formula but different molecular struc-

ture. Therefore, CH_3OCH_3 and CH_3CH_2OH differ from each other based on their molecular structure. If two compounds have the same molecular formula, they must have the same number of atoms (3) and the same formula mass (4).

50. **4** If a hydrogen is replaced by an —OH functional group, then the molecule becomes an alcohol. Alcohol's characteristic functional group is —OH.

Functional group	Type of molecule
—COOH	Carboxylic acid
—COH	Aldehyde
—CO—	Ketone
—COO—	Ester
—OH	Alcohol

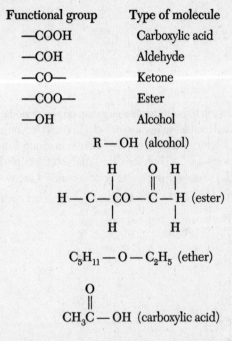

R — OH (alcohol)

C_5H_{11} — O — C_2H_5 (ether)

CH_3C — OH (carboxylic acid)

51. **3** All organic hydrocarbons with the general formula of C_nH_{2n-2} are called *alkynes*. Alkynes are organic molecules with a triple bond in their structure. The only structure that represents an alkyne is choice 3.

52. **4** The valence electrons are located on the outermost principal energy level of an atom. Because the electron configuration of carbon is $1s^2 2s^2 2p^2$ (recall that carbon's atomic number is 6), the total number of valence electrons must be 4. The highest principal energy level of carbon is the second level, or $2s^2 2p^2$. Therefore, the correct answer is choice 4.

53. 1 As we move down a group, an element's nonmetallic characteristic decreases. We can verify this by referring to its electronegative values. The greater the electronegative value, the greater the nonmetallic characteristics. According to Reference Table K, electronegative value decreases from top to bottom within a group.

Group 17 elements	Electronegativity
F	4.0
Cl	3.2
Br	2.9
I	2.7
At	2.2

54. 2 Moving from top to bottom within a group on the periodic table, the elements have increasing numbers of electrons. To accommodate additional electrons, each successive element in group 1 must add principal energy levels. Therefore, the number of occupied principal energy levels increases from top to bottom within a group.

Group 1 elements	Highest energy level
H	1
Li	2
Na	3
K	4
Rb	5
Cs	6
Fr	7

55. 3 Isotopes of the same element differ from each other based on the number of neutrons each isotope possesses. Therefore, a change in the mass number (mass number = protons + neutrons) in hydrogen isotopes simply indicates a change in the number of neutrons, and the number of protons remains unaffected.

56. 1 $Cu(NO_3)_2$ is a salt. When a salt is dissolved in water, a strong acid and a weak base are produced.

$$Cu(NO_3)_2 + 2H_2O \rightarrow 2HNO_3 + Cu(OH)_2$$
$$\text{strong acid} \quad \text{weak base}$$

Because a strong acid is produced, the solution becomes more acidic. In other words, the pH of the solution decreases.

PART 2
Group 1—Matter and Energy

57. 1 Sublimation is the process through which a solid is changed into a gas without first going through the liquid phase.

Type of phase change	Name of process
Solid to gas	Sublimation
Gas to solid	Deposition
Gas to liquid	Condensation
Liquid to gas	Boiling (evaporation)
Liquid to solid	Freezing
Solid to liquid	Melting

58. 3 Atomic mass (choice 1) and atomic radius (choice 2) are fixed characteristics of an atom. No physical force (e.g., temperature) can change the fundamental structure of atoms. The melting point (choice 4) varies according to pressure and not temperature. On the other hand, when the temperature of a liquid increases, its vapor pressure increases as well. Vapor pressure is the pressure created by evaporating liquid molecules. As temperature increases, more liquid molecules evaporate, thus leading to a higher vapor pressure.

59. 4 The question asks for a chemical property of the element iodine (I_2). The first three choices are all physical properties. Chemical properties define the nature of reactions an element undergoes. Choice 4 is the correct answer because it is the only choice that defines a chemical property.

$$I_2(s) + H_2(s) \rightarrow 2HI(g)$$

60. 2 Particles in all three phases of matter vibrate at various frequencies. However, vibrating particles are not necessarily mobile. Mobile in this

case means able to perform translational movement. Depending on the phase of matter, vibrating particles can be mobile or stationary. In the case of solids, vibrating particles are stationary. On the other hand, particles are relatively mobile in liquid phase and highly mobile in gaseous phase.

61. 1 According to Charles' law, the volume of a gas is directly proportional to the temperature (measure in K) at constant pressure. As the temperature increases, the gas expands in volume.

$V_1/T_1 = V_2/T_2 = $ constant

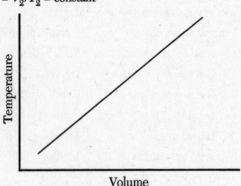

Temperature (K) and volume are
directly related to each other
(Charles' Law)

Group 2—Atomic Structure

62. 3 The number at the top left corner of the box for each element in the periodic table refers to the atomic mass. Atomic mass is the sum of the total number of protons and neutrons present in an atom's nucleus. Because ^{35}Cl and ^{37}Cl are isotopes of the same element, they must differ from each other with respect to their number of neutrons. Isotopes have the same number of protons.

63. 1 The total number of electrons present in a neutral atom is the same as the total number of protons. In the atom $^{59}_{27}Co$, there are 27 protons (atomic number). Because the overall charge on this atom is zero (no charge), the number of electrons must be the same as the number of protons. Therefore, the total number of electrons is 27.

64. 2 The atom with the highest number of protons has the greatest nuclear charge. Therefore, the best way to answer this question is to compare the atomic numbers of the given elements in the four answer choices. Based on the periodic table, it is evident that argon (Ar) has the highest atomic number of the four elements given.

Element	Atomic number
Al	13
Ar	18
Na	11
Si	14

65. 3 The element with the atomic number 18 is argon (Ar). That means Ar has 18 protons and electrons. The electron configuration of Ar is

$1s^2 2s^2 2p^6 3s^2 3p^6$

Based on the electron configuration, the principal quantum number of its outermost electrons is 3.

66. 2 According to Reference Table H, the half-life of ^{222}Rn is 3.82 days. Therefore, after 19.1 days, the following amount of the original 160 mg of ^{222}Rn remains:

Amount of ^{222}Rn remaining = original mass $\times (\frac{1}{2})^n$,
where n = number of half-lives passed, and
n = 19.1 days/3.82 days = 5 half-lives have passed.

$$
\begin{aligned}
\text{Amount of } ^{222}\text{Rn remaining} &= 160 \text{ mg} \times (\tfrac{1}{2})^5 \\
&= 160 \text{ mg} \times \tfrac{1}{32} \\
&= 5.0 \text{ mg of } ^{222}\text{Rn remaining}
\end{aligned}
$$

Group 3—Bonding

67. 3 Covalent bonds are present between two nonmetals. In a covalent bond, the nonmetals share electrons. However, the electrons can be shared equally or unequally. If electrons are equally shared, then the bond is called a *nonpolar covalent bond*. If electrons are unequally shared, then the bond is called a *polar covalent bond*. A nonpolar covalent bond is formed between two identical nonmetals because their electronegativities are identical as well. Nonmetals with identical electronegativities have the same affinity for electrons. As a result,

shared electrons spend equal time around the two nuclei of the non-metals. The choice that correctly characterizes the nature of a nonpolar covalent bond is choice 3.

Bond type	Bond description	Where is it found
Metallic	Electrons are shared by multiple ions	All metals
Ionic	Electrons are transferred from metals to nonmetals	All salts
Nonpolar Covalent	Electrons are equally shared by nonmetals	Diatomic molecules (F_2, Cl_2)
Polar Covalent	Electrons are unequally shared by nonmetals	Most molecules that are covalent in nature

68. 4 Empirical formula is the lowest possible whole number ratio of elements in a molecular compound.

Name	Empirical formula	Molecular formula
Acetylene	CH	C_2H_2
Ethene	CH_2	C_2H_4
Ethane	CH_3	C_2H_6
Methane	CH_4	CH_4

69. 3 When balancing the combustion of hydrocarbons, first balance the carbon and then the hydrogen. Save the oxygen for last. Remember that it is easiest to balance all chemical equations by saving for last the elements that occur in the greatest number of compounds. To balance the given equation, put an 8 in front of the CO_2 and an 8 in front of the H_2O to balance carbon and hydrogen. Now there are 24 oxygen on the right side of the equation: 16 from carbon dioxide and 8 from water. By putting a 12 in front of O_2, the equation is balanced.

$$C_8H_{16} + 12O_2 \rightarrow 8CO_2 + 8H_2O$$

According to the balanced equation, the smallest whole number coefficient of O_2 is 12.

70. 2 A coordinate covalent bond is formed when the pair of electrons shared in a covalent bond is supplied by just one atom. The H^+ ion lacks any electrons. That means an atom that can bond with H^+ must supply a pair of available electrons to form a bond. In the answer choices given, only choice 2 ($H:^-$) has two electrons available to form a bond. $H\cdot$ (choice 1) has a single electron, H^+ (choice 3) has no electrons, and $H:H$ (choice 4) indicates a regular covalently bonded hydrogen (electrons are already being shared by the two hydrogens).

71. 4 Molecule-ion attractions only occur in a solution of an ionic compound in a molecular solvent. Choices 1, 2, and 3 all indicate pure KCl. Choice 4 indicates an aqueous solution of potassium chloride, which contains ions (K^+ and Cl^-) and molecules (H_2O). The ions and the water molecules attract each other. These interparticle attractions are called *molecule-ion attractions*.

Group 4—Periodic Table

72. 2 Elements that lose electrons during formation of ions have atoms with radii larger than the radii of their ions. In positive ions, the positive charge of the nucleus is greater than the negative charge of the electrons, so the electrons are pulled in tighter in the ion than in the neutral atom. The ionic radii of positive ions is smaller than the atomic radii of the corresponding neutral atoms. For atoms that form negative ions, there is a greater negative charge than positive charge in the nucleus, and so it is more difficult for the nucleus to pull in the electrons as compared with the neutral atom. The ionic radius of negative ions is greater than the atomic radii of the corresponding neutral atoms. Only metals lose electrons when they form ions and have atomic radii larger than their ionic radii. Of the four choices given, the only choice that represents a metal is calcium (Ca).

73. 3 The periodic table is arranged according to elements' atomic numbers. Scientists have noted that the chemical properties of elements repeat after regular increases in atomic number. Increases in atomic number according to the following pattern lead to elements with simi-

lar chemical properties: 8, 8 and then 18, and 18. For example, all of the noble gases, which are similar in their nonreactivity, have atomic numbers 2, 10, 18, 36, and 54.

$2 + 8 = 10$
$10 + 8 = 18$
$18 + 18 = 36$
$36 + 18 = 54$

Furthermore, elements of similar chemical properties are placed into groups. Masses, weights, and radii are physical properties of atoms.

74. **2** Metals are the best conductors of electricity. Cu(s) is the only metal of the four choices given. NaCl (choice 1) is an ionic compound, and H_2O (choice 3) and Br_2 (choice 4) are molecular compounds.

75. **4** Iodine is solid at STP.

Halogen	Phase at STP
Fluorine	Gas
Chlorine	Gas
Bromine	Liquid
Iodine	Solid

Iodine is also the largest halogen, and we would expect it to be the most likely to be solid. It exhibits the greatest dispersion forces.

76. **1** When nonmetals and metals react, they form ionic compounds. During an ionic reaction, electrons are transferred from metals to nonmetals. As a result, metals lose electrons (oxidation) and nonmetals gain electrons (reduction). Therefore, nonmetals react with atoms of metals by gaining electrons to form ionic compounds.

Group 5—Mathematics of Chemistry

77. **3** Start solving this problem by assuming that there is a 100-g sample of the compound. Because the compound is 2.8% boron (B) and 97% iodine (I), we assume that there are 2.8 g of B and 97 g of I in 100 g of the compound. Next, convert 2.8 g of B and 97 g of I to moles by dividing their masses by their respective molar masses.

$$B = \frac{2.8 \text{ g}}{10.8 \text{ g}} = 0.259 \text{ mol}$$

$$I = \frac{97 \text{ g}}{127 \text{ g}} = 0.764 \text{ mol}$$

Divide both calculated moles with the lower of the two quantities to obtain the smallest whole number ratio possible.

B = 0.259 mol/0.259 mol = 1
I = 0.764 mol/0.259 mol = 3

Based on the mole ratio, the empirical formula must be BI_3.

78. **3** Use the combined gas formula to solve the problem:

$$\frac{P_1 V_1}{T_1} = \frac{P_2 V_2}{T_2}$$

$$V_2 = \frac{P_1 V_1 T_2}{P_2 T_1}$$

$V_1 = 2 \text{ L}$ $T_1 = 323 \text{ K}$ $P_1 = 3 \text{ atm}$
$V_2 = ?$ $T_2 = 273 \text{ K}$ $P_2 = 1 \text{ atm}$

$$V_2 = 2 \text{ L} \times \frac{273 \text{ K}}{323 \text{ K}} \times \frac{3 \text{ atm}}{1 \text{ atm}}$$

79. **2** Use the following equations to solve this problem:

Equation 1: density = mass/volume
Equation 2: mass = volume × density
             (cross-multiply equation 1 to obtain equation 2)

Assume there is 1 mole of the unknown gas. Because 1 mole of all gases occupies 22.4 liters at STP, we can use this value as our volume.

Mass = volume × density
        = 22.4 L × 2.05 g/L
        = 45.9 g

Because 1 mole of the unknown gas must have a mass of 45.9 g, 45.9 g/mol is the molar mass of the unknown. Of the four choices given, only NO_2 has a molar mass of 46.

80. 3 Use the following formula to solve this problem:

Heat absorbed in calories (q) = mass of water × heat of vaporization (ΔH_v)

$$q = m \times \Delta H_v$$
$$q = 10.0 \text{ g} \times 539.4 \text{ cal/g} \ (\Delta H_v = 539.4 \text{ cal/g},$$
according to Reference Table A)
$$q = 5{,}390 \text{ calories}$$

81. 3 The formula for freezing point depression is

Freezing point depressed = molality × freezing point constant

$$\Delta T_f = m \times K_f$$

5.58°C = molality × 1.86°C kg/mol (the freezing point constant can be obtained from Reference Table A)

Molality = 5.58°C/1.86°C kg/mol = 3.00 kg/mol

Molality = moles of solute/kg solvent

Moles of solute = molality/kg solvent

$$= \frac{3.00 \text{ mol/kg}}{1 \text{ kg}} \qquad (1{,}000 \text{ g} = 1 \text{ kg})$$

$$= 3 \text{ mol}$$

Group 6—Kinetics and Equilibrium

82. 3 The compound that undergoes a spontaneous formation reaction must have a negative free energy (ΔG_f) regardless of whether ΔH_f is positive or negative. Based on Reference Table G, only choice 3 (ICl) has a negative ΔG_f. Therefore, formation of ICl(g) must be a spontaneous reaction.

Compound	ΔG_f (kcal/mol)
C_2H_4	+16.3
C_2H_2	+50.0
ICl	–1.3
HI	+0.4

83. 2 This problem can be solved by determining how the equilibrium of this reaction is affected by the addition of Cl⁻ ions. According to the question, the AgCl dissociates to form Ag⁺ and Cl⁻ ions in solution. The amount of Ag⁺ and Cl⁻ ions formed in solution is determined by

AgCl's solubility constant (K_{sp}). Remember, any equilibrium constant must remain constant regardless of any stress experienced by the reaction. And according to the Le Châtelier principle, if a stress is applied to an equilibrium, the system readjusts to counterbalance the stress.

$$K_{sp} = [Ag^+] \times [Cl^-] = 1.8 \times 10^{-10} = \text{constant}$$

If Cl^- is added, thus increasing the concentration of Cl^- in solution, the concentration of Ag^+ must decrease to compensate for the increased Cl^-. In other words, if the concentration of one of the ions increases, the concentration of the other ion must decrease to maintain the same K_{sp} value. The concentration of Ag^+ decreases as excess Cl^- reacts with Ag^+ to form AgCl(s). The amount of AgCl(s) is therefore increased.

84. **3** The expression $\Delta H - T\Delta S$ is equal to ΔG. ΔG is also called the *free energy of a reaction*.

ΔH = enthalpy, or heat of reaction
ΔS = entropy, or randomness of reaction
T = temperature in K
Free energy = change in enthalpy – temperature × change in entropy

85. **1** Precipitation reactions have the fastest reaction rate. In the four choices given, only the reaction in choice 1 is a precipitation reaction.

$$Pb^{2+}(aq) + S^{2-}(aq) \rightarrow PbS(s)$$

86. **4** When a compound reaches its saturation point, the rate at which ions of the compound dissolve into solution is the same as the rate at which already dissolved ions recombine to form the original compound. How concentrated or diluted the solution is depends on the solubility of the compound and not on whether the compound has reached its saturation point.

Group 7—Acids and Bases

87. **1** An Arrhenius base is a compound that can release OH^- into water. Therefore, OH^- is the only negative ion present in the solution when an Arrhenius base is placed in water.

For example, $KOH(s) \overset{H_2O}{\rightarrow} K^+(aq) + OH^-(aq)$

88. 4 Litmus paper changes from red to blue if there is a base present. Of the four choices given, only NaOH (choice 4) is a base. HCl (choice 1) is an acid, NaCl (choice 2) is a salt, and CH_3OH (choice 3) is an alcohol.

89. 4 A Brönsted base is a compound that can accept an acidic proton. An acidic proton is also characterized as the H^+ ion. Therefore, when an Cl^- ion combines with an H^+ ion, it acts as a Brönsted base.

90. 3 An amphoteric substance can act as an acid by donating an H^+ ion or can act as a base by receiving an H^+ ion. In choice 3, water acts as both an acid and a base. One of the water molecules gives away an H^+ ion to form an OH^- ion. The other water molecule receives a proton to form H_3O^+.

$$H_2O + H_2O \rightarrow H_3O^+ + OH^-$$

91. 1 Use the following formula to solve this problem:

$$\text{Volume}_{(acid)} \times \text{molarity}_{(acid)} = \text{volume}_{(base)} \times \text{molarity}_{(base)}$$

$$\text{Molarity}_{(acid)} = \frac{\text{volume}_{(base)} \times \text{molarity}_{(base)}}{\text{volume}_{(acid)}}$$
$$= (17.3 \text{ ml})(0.250 \text{ M})/37.6 \text{ ml}$$
$$= 0.115 \text{ M}$$

Group 8—Redox and Electrochemistry

92. 3 When the switch is closed in an electrochemical cell, electrons flow from the electrode where oxidation takes place to the electrode where reduction occurs. In this question, oxidation occurs at the Zn electrode (A) and reduction occurs at the Cu electrode (D). The salt bridge is created to complete the circuit inside an electrochemical cell. The salt bridge allows excess ions from both solutions to move in either direction to compensate for the electrical imbalance that is created as electrons migrate from electrode A to electrode D. Negative SO_4^{2-} ions migrate from C to B because an excess of Zn^{2+} ions is produced at the anode, and positive Zn^{2+} ions migrate from B to C because the positive ion concentration in the copper sulfate solution decreases as the Cu^{2+} ions are reduced (solutions must be electrically neutral). The correct answer is choice 3 because it describes the function of the salt bridge accurately.

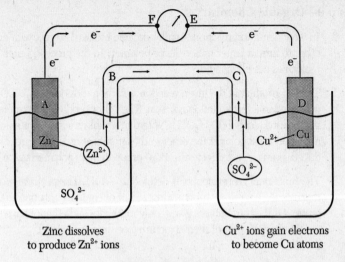

Zinc dissolves
to produce Zn^{2+} ions

Cu^{2+} ions gain electrons
to become Cu atoms

Electrochemical cell

93. **2** As described in question 92, electrons flow from the Zn electrode (A) to the Cu electrode (D). On their way to electrode D from electrode A, electrons also pass through the voltmeter (F) and the switch (E). Therefore, the correct electron flow pathway is $A \rightarrow F \rightarrow E \rightarrow D$ (choice 2). Electrons do not flow through the solution, only through the metal electrodes and metal wire.

94. **4** Reference Table N lists the reduction potential of various elements. To react with H^+, an ion must have a reduction potential that is lower than that of H^+. The reduction potential of H_2 is 0.00. Of the answer choices given, only Mg^{2+} has a reduction potential that is lower than that of H_2. The reduction potential of Mg^{2+} is -2.37.

95. **1** The standard reduction potential of $Cu^{+2} + 2e^- \rightarrow Cu(s)$ can be found in Reference Table N. According to the table, the standard electrode potential of the half-reaction mentioned above is $+0.34$ V.

96. **1** In an electrochemical cell, the anode is the site of the oxidation reaction. In an oxidation reaction, an element loses electrons and its oxidation number increases. In the reaction $Zn(s) + Pb^{2+}(aq) \rightarrow Zn^{2+}(aq) + Pb(s)$, Zn undergoes oxidation to become Zn^{2+}. Therefore, Zn must be oxidized at the anode.

Group 9—Organic Chemistry

97. **2** In a condensation reaction, smaller organic molecules react with each other to form a larger molecule, or polymer. In the process, water molecules are eliminated.

98. **1** The key product in the given reaction is carbon dioxide (CO_2). CO_2 is produced in a fermentation reaction. The other product of fermentation is ethyl alcohol (C_2H_5OH). Note that a molecule that is nonpolar at one end and polar at the other end (saponification) is not produced, nor is an ester (esterification) or a polymer (polymerization).

99. **3** The functional group alcohol is defined as —OH. The prefix *mono* means one. Therefore, monohydroxy alcohol contains only one —OH group. Of the four choices given, only methanol and ethanol have one —OH group as part of their structure (see diagram below).

$$
\begin{array}{ccc}
& \text{H} \quad \text{H} & \\
& | \qquad | & \\
\text{H} - & \text{C} - \text{C} - \text{H} & \quad \text{Ethylene Glycol} \\
& | \qquad | & \\
& \text{OH} \quad \text{OH} &
\end{array}
$$

$$
\begin{array}{ccc}
& \text{H} \quad \text{H} & \\
& | \qquad | & \\
\text{H} - & \text{C} - \text{C} - \text{OH} & \quad \text{Ethanol} \\
& | \qquad | & \\
& \text{H} \quad \text{H} &
\end{array}
$$

$$
\begin{array}{ccc}
\text{OH} \;\; \text{OH} \;\; \text{OH} & \\
| \qquad | \qquad | & \\
\text{H} - \text{C} - \text{C} - \text{C} - \text{H} & \quad \text{Glycerol} \\
| \qquad | \qquad | & \\
\text{H} \;\;\; \text{H} \;\;\; \text{H} &
\end{array}
$$

$$
\begin{array}{ccc}
\text{OH} & \\
| & \\
\text{H} - \text{C} - \text{H} & \quad \text{Methanol} \\
| & \\
\text{H} &
\end{array}
$$

100. **4** The structure represents an ether.

$$
\begin{array}{c}
\text{H} \quad \text{H} \qquad\qquad \text{H} \quad \text{H} \\
| \qquad | \qquad\qquad | \qquad | \\
\text{H} - \text{C} - \text{C} - \text{O} - \text{C} - \text{C} - \text{H} \\
| \qquad | \qquad\qquad | \qquad | \\
\underbrace{\text{H} \quad \text{H}}_{\text{Ethyl}} \quad\;\; \underbrace{\text{H} \quad \text{H}}_{\text{Ethyl}} \\
\text{Ether}
\end{array}
$$

101. 3 2-Chlorobutane must have a chlorine (Cl) atom attached to the second carbon of the four-carbon-long butane molecule. In choice 3, a single chlorine is attached to the second carbon of the butane. Choice 2 is incorrect because the chlorine atom is attached to the first carbon. Choices 1 and 4 are incorrect because two chlorine atoms are attached to the butane molecule.

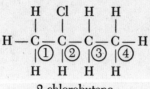

2-chlorobutane

Group 10—Applications of Chemical Principles

102. 1 A lead-acid battery produces electricity during discharge. Discharge is the forward reaction.

$$Pb + PbO_2 + 2H_2SO_4 \rightarrow 2PbSO_4 + 2H_2O$$

When an element undergoes oxidation, its oxidation number increases. In the given reaction, Pb^0 oxidized to Pb^{+2} (Pb's charge is +2 in $PbSO_4$). In addition, Pb^{+4} from PbO_2 is reduced to Pb^{+2} in the $PbSO_4$.

103. 1 Hydrocarbons with the lowest molecular weight also have the lowest boiling point. Of the four answer choices given, the molecule with the lowest number of carbon and hydrogen has the lowest molecular weight and, therefore, it also has the lowest boiling point. Small molecules tend to have weaker van der Waals forces between them. As a result, the boiling point is comparatively lower.

104. 3 Petroleum is one of the most valuable commodities in the world. Not only is petroleum the starting point for many fuels, but it also is used to synthesize various types of synthetic textiles and plastic products, such as nylon, polyester, and polyvinyl chloride.

105. 1 The function of all catalysts is to increase the rate at which reactions take place. Reaction rate is increased by lowering the activation energy. The lowering of the activation energy is accomplished by catalysts. Catalysts affect no other aspects of reactions.

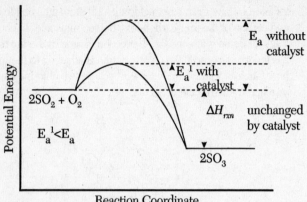

E_a without catalyst

E_a^1 with catalyst

ΔH_{rxn} unchanged by catalyst

$2SO_2 + O_2$

$E_a^1 < E_a$

$2SO_3$

Reaction Coordinate

Potential Energy

106. 4 In the reaction

$$2Al + Cr_2O_3 \rightarrow Al_2O_3 + 2Cr,$$

chromium (Cr) undergoes reduction because its oxidation number changes from Cr^{+3} to Cr^0. The aluminum must have undergone oxidation in the same reaction. The species that undergoes an oxidation so that another species can undergo a reduction is labeled the reducing agent.

Group 11—Nuclear Chemistry

107. 2 The diagram illustrates a larger nucleus breaking up into two smaller nuclei. Such a nuclear deformation is called a *fission reaction*. In a fission reaction, energy is always released. Therefore, fission reactions are exothermic reactions.

108. 4 Moderators slow neutrons down by absorbing them so that they cannot participate in further fission reactions. A moderator has the ability to intervene in a regular nuclear reaction. Moderators are often used to slow down nuclear reactors to increase their life and efficiency. Heavy water and graphite are two examples of moderators.

109. 3 Only charged particles (positive or negative) can be accelerated in a magnetic field. A particle that has no charge or is neutral in nature cannot be accelerated in a magnetic field. Of the four answer choices given, only a neutron lacks a charge. In other words, it is neutral in

nature. Therefore, neutrons do not accelerate when exposed to a magnetic field. Alpha particles and protons are positively charged, and beta particles are negatively charged.

110. **2** Thyroid hormones contain iodine as part of their structure. Iodine is unique to the thyroid gland and the hormones it produces. Therefore, thyroid disorders would be best diagnosed by using a radioisotope of iodine or iodine-131.

111. **1** In a nuclear reaction, the total mass (atomic mass) as well as atomic numbers must be equal on both sides of the reaction.

$$
\begin{aligned}
\text{Left side} \quad &= \quad \text{Right side} \\
0 + 92 \quad &= \quad 56 + x + 3(0) \\
92 \quad &= \quad 56 + x \\
x \quad &= \quad 36
\end{aligned}
$$

Based on these calculations, the unknown (x) has an atomic number of 36.

Group 12—Laboratory Activities

112. **4** The test tube holder is represented by the figure in choice 4. Choice 1 is a hose clamp, choice 2 are tongs, and choice 3 is a ring-stand clamp.

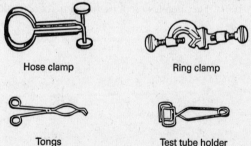

Hose clamp

Ring clamp

Tongs

Test tube holder

113. **2** Percent error can be calculated by using the following formula:

$$
\% \text{ error} = \frac{\text{experimental value} - \text{accepted value} \times 100\%}{\text{accepted value}}
$$

$$
= \frac{2.25 \text{ L}}{22.4 \text{ L}} \times 100 = 10\%
$$

114. 4   To solve this problem, the following formula has to be used:

Density = mass/volume

Based on this formula, we need to know the mass and volume of the object to calculate its density.

Volume of the object = 27.8 ml – 21.2 ml = 6.6 ml
Mass of the object = 22.4 g
Density = 22.4 g/6.6 ml
= 3.4 g/ml

115. 4   Electricity flows only if the solution contains ions or charged particles. KCl(aq) represents a solution that has K^+ and Cl^- dissolved in water. Therefore, KCl(aq) makes the lightbulb grow brightly. Choices 1 and 2 are examples of organic molecules. Molecules of this type do not form charged particles. Choice 3 is an example of an ionic solid. Unless ionic compounds are dissolved in solution or are molten, they do not conduct electricity.

116. 3   All endothermic processes, including reactions, absorb energy or heat. When energy or heat is absorbed during an endothermic process, the surrounding temperature decreases. The decrease in temperature is a hallmark of an endothermic process. Choice 3 is the correct answer because it accurately describes the nature of an endothermic process.

EXAMINATION
JUNE 1994

PART 1: *Answer all 56 questions in this part.* [65]

DIRECTIONS (**1–56**): *For each statement or question, select the word or expression that, of those given, best completes the statement or answers the question. Record your answer on the separate answer sheet provided.*

---

1 The amount of energy needed to change a given mass of ice to water at constant temperature is called the heat of
1 condensation   3 fusion
2 crystallization   4 formation

2 Which equation represents the phase change called sublimation?
(1) $CO_2(s) \rightarrow CO_2(g)$
(2) $H_2O(s) \rightarrow H_2O(\ell)$
(3) $H_2O(\ell) \rightarrow H_2O(g)$
(4) $NaCl(\ell) \rightarrow NaCl(s)$

3 Which substance can *not* be decomposed into simpler substances?
1 ammonia   3 methane
2 aluminum   4 methanol

4 In which sample are the patterns arranged in a regular geometric pattern?
(1) $HCl(\ell)$ (3) $N_2(g)$
(2) $NaCl(aq)$ (4) $I_2(s)$

5 How many calories are equivalent to 35 kilocalories?
   (1) 0.035 calorie   (3) 3,500 calories
   (2) 0.35 calorie   (4) 35,000 calories

6 How does the ground state electron configuration of the
   hydrogen atom differ from that of a ground state helium
   atom?
   1 Hydrogen has one electron in a higher energy level.
   2 Hydrogen has two electrons in a lower energy level.
   3 Hydrogen contains a half-filled orbital.
   4 Hydrogen contains a completely filled orbital.

7 Which type of radiation would be attracted to the positive
   electrode in an electric field?
   (1) $_{-1}^{0}e$ (3) $_{2}^{3}He$
   (2) $_{1}^{1}H$ (4) $_{0}^{1}n$

8 Which electron configuration represents an atom in an
   excited state?
   (1) $1s^2 2s^2$ (3) $1s^2 2s^2 2p^6$
   (2) $1s^2 2s^2 3s^1$ (4) $1s^2 2s^2 2p^6 3s^1$

9 Which electron transition represents the release of energy?
   (1) $1s$ to $3p$ (3) $3p$ to $1s$
   (2) $2s$ to $2p$ (4) $2p$ to $3s$

10 The atomic number of any atom is equal to the number of
   1 neutrons in the atom, only
   2 protons in the atom, only
   3 neutrons plus protons in the atom
   4 protons plus electrons in the atom

11 The mass of an electron is approximately $\frac{1}{1836}$ times the mass of
   (1) ^1_1H (3) ^3_1H
   (2) ^2_1H (4) ^4_2He

12 Which nuclear reaction is classified as alpha decay?
   (1) $^{14}_6C \rightarrow {}^{14}_7N + {}^0_{-1}e$
   (2) $^{42}_{19}K \rightarrow {}^{42}_{20}Ca + {}^0_{-1}e$
   (3) $^{226}_{88}Ra \rightarrow {}^{222}_{86}Rn + {}^4_2He$
   (4) $^3_1H \rightarrow {}^0_{-1}e + {}^3_2He$

13 Which diagram correctly shows the relationship between electronegativity and atomic number for the elements of Period 3?

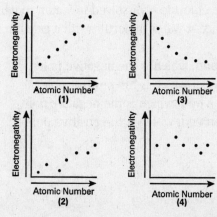

14 Which statement is true concerning the reaction
$N(g) + N(g) \rightarrow N_2(g) + energy$?
1 A bond is broken and energy is absorbed.
2 A bond is broken and energy is released.
3 A bond is formed and energy is absorbed.
4 A bond is formed and energy is released.

15 Hydrogen bonding is strongest between molecules of
(1) H_2S (3) H_2Se
(2) H_2O (4) H_2Te

16 A chemical formula is an expression used to represent
1 mixtures, only
2 elements, only
3 compounds, only
4 compounds and elements

17 Which calcium chloride is dissolved in water, to which
end of the adjacent water molecules will a calcium ion
be attracted?
1 the oxygen end, which is the negative pole
2 the oxygen end, which is the positive pole
3 the hydrogen end, which is the negative pole
4 the hydrogen end, which is the positive pole

18 How do the chemical properties of the atom and the Na⁺ ion compare?
   1 They are the same because each has the same atomic number.
   2 They are the same because each has the same electron configuration.
   3 They are different because each has a different atomic number.
   4 They are different because each has a different electron configuration.

19 Which element in Group 16 (VIA) has *no* stable isotopes?
   1 sulfur   3 tellurium
   2 selenium   4 polonium

20 Which element is a member of the halogen family?
   (1) K   (3) I
   (2) B   (4) S

21 Most nonmetals have the properties of
   1 high ionization energy and poor electrical conductivity
   2 high ionization energy and good electrical conductivity
   3 low ionization energy and poor electrical conductivity
   4 low ionization energy and good electrical conductivity

22 The metalloids that are included in Group 15 (VA) are
    antimony (Sb) and
    (1) N   (3) As
    (2) P   (4) Bi

23 Which of the following atoms has the largest atomic
    radius?
    (1) Na   (3) Mg
    (2) K   (4) Ca

24 In which group does each element have a total of four
    electrons in the outermost principal energy level?
    (1) 1 (IA)   (3) 16 (VIA)
    (2) 18 (0)   (4) 14 (IVA)

25 Which properties are characteristic of the Group 1 (IA)
    metals?
    1 high reactivity and the formation of stable compounds
    2 high reactivity and the formation of unstable compounds
    3 low reactivity and the formation of stable compounds
    4 low reactivity and the formation of unstable compounds

26 Which quantity is equivalent to 39 grams of LiF?
    (1) 1.0 mole   (3) 0.50 mole
    (2) 2.0 moles   (4) 1.5 moles

27 Which quantity represents 0.500 mole at STP?
    (1) 22.4 liters of nitrogen
    (2) 11.2 liters of oxygen
    (3) 32.0 grams of oxygen
    (4) 28.0 grams of nitrogen

28 When 0.50 liter of a 12 M solution is diluted to 1.0 liter, the molarity of the new solution is
(1) 6.0 M   (3) 12 M
(2) 2.4 M   (4) 24 M

29 What is the percent by mass of oxygen in Fe_2O_3 (formula mass = 160)?
(1) 16%   (3) 56%
(2) 30.%   (4) 70.%

30 Given the equation:

$$6CO_2(g) + 6H_2O(\ell) \rightarrow C_6H_{12}O_6(s) + 6O_2(g)$$

What is the *minimum* number of liters of $CO_2(g)$, measured at STP, needed to produce 32.0 grams of oxygen?
(1) 264 L
(2) 192 L
(3) 32.0 L
(4) 22.4 L

31 Given the reaction:

$$Mg(s) + 2HCl(aq) \rightarrow MgCl_2(aq) + H_2(g)$$

The reaction occurs more rapidly when a 10-gram sample of Mg is powdered, rather than in one piece, because powdered Mg has
1 less surface area
2 more surface area
3 a lower potential energy
4 a higher potential energy

32 According to Reference Table *I*, in which reaction do the products have a higher energy content than the reactants?

(1) $CH_4(g) + 2O_2(g) \rightarrow CO_2(g) + 2H_2O(\ell)$

(2) $CH_3OH(\ell) + \frac{3}{2}O_2(g) \rightarrow CO_2(g) + 2H_2O(\ell)$

(3) $NH_4Cl(s) \xrightarrow{\text{H}_2\text{O}} NH_4^+(aq) + Cl^-(aq)$

(4) $NaOH(s) \xrightarrow{\text{H}_2\text{O}} Na^+(aq) + OH^-(aq)$

33 Which equation correctly represents the free energy change in a chemical reaction?

(1) $\Delta G = \Delta H + T\Delta S$

(2) $\Delta G = \Delta H - T\Delta S$

(3) $\Delta G = \Delta T - \Delta H \Delta S$

(4) $\Delta G = \Delta S - T\Delta H$

34 Adding a catalyst to a chemical reaction will

1 lower the activation energy needed

2 lower the potential energy of the reactants

3 increase the activation energy needed

4 increase the potential energy of the reactants

35 Under which conditions are gases most soluble in water?

1 high pressure and high temperature

2 high pressure and low temperature

3 low pressure and high temperature

4 low pressure and low temperature

36 A solution of a base differs from a solution of an acid in that the solution of a base

1 is able to conduct electricity

2 is able to cause an indicator color change

3 has a greater $[H_3O^+]$

4 has a greater $[OH^-]$

37 Given the reaction:

$$HCl(g) + H_2O(\ell) \rightarrow H_3O^+(aq) + Cl^-(aq)$$

Which reactant acted as a Brönsted-Lowry acid?
(1) $HCl(g)$, because it reacted with chloride ions
(2) $H_2O(\ell)$, because it produced hydronium ions
(3) $HCl(g)$, because it donated protons
(4) $H_2O(\ell)$, because it accepted protons

38 Which of the following aqueous solutions is the *poorest* conductor of electricity? [Refer to Reference Table *L*]
(1) 0.1 M H_2S (3) 0.1 M HNO_2
(2) 0.1 M HF   (4) 0.1 M HNO_3

39 According to the Arrhenius theory, the acidic property of an aqueous solution is due to an excess of
(1) H_2 (3) H_2O
(2) H^+ (4) OH^-

40 Which pH value indicates the most basic solution?
(1) 7   (3) 3
(2) 8   (4) 11

41 A 3.0-milliliter sample of HNO_3 solution is exactly neutralized by 6.0 milliliters of 0.50 M KOH. What is the molarity of the HNO_3 sample?
(1) 1.0 M   (3) 3.0 M
(2) 0.50 M   (4) 1.5 M

42 As 100 milliliters of 0.10 molar KOH is added to 100 milliliters of 0.10 molar HCl at 298 K, the pH of the resulting solution will
1 decrease to 3   3 increase to 7
2 decrease to 14   4 increase to 13

43 What is the oxidation number of oxygen in HSO_4^-?
(1) +1   (3) +6
(2) –2   (4) –4

44 Which half-reaction correctly represents a reduction reaction?
(1) $Sn^0 + 2e^- \rightarrow Sn^{2+}$
(2) $Na^0 + e^- \rightarrow Na^+$
(3) $Li^0 + e^- \rightarrow Li^+$
(4) $Br_2^0 + 2e^- \rightarrow 2Br^-$

45 Given the reaction:

$$2Na + 2H_2O \rightarrow 2Na^+ + 2OH^- + H_2$$

Which substance is oxidized?
(1) H_2 (3) Na
(2) H^+ (4) Na^+

46 Which change occurs when an Sn^{2+} ion is oxidized?
1 Two electrons are lost.
2 Two electrons are gained.
3 Two protons are lost.
4 Two protons and gained.

47 Based on Reference Table N, which of the following elements is the strongest reducing agent?
(1) Fe
(3) Cu
(2) Sr
(4) Cr

48 An electrochemical cell that generates electricity contains half-cells that produce
1 oxidation half-reactions, only
2 reduction half-reactions, only
3 spontaneous redox reactions
4 nonspontaneous redox reactions

49 Which structural formula represents a molecule of butane?

(1)
$$H-\underset{\underset{H}{|}}{C}=\underset{\underset{H}{|}}{C}-\underset{\underset{H}{|}}{C}=\underset{\underset{H}{|}}{C}-H$$

(2)
$$H-\underset{\underset{H}{|}}{\overset{\overset{H}{|}}{C}}-\overset{\overset{H}{|}}{C}=\overset{\overset{H}{|}}{C}-\underset{\underset{H}{|}}{\overset{\overset{H}{|}}{C}}-H$$

(3)
$$H-\underset{\underset{H}{|}}{\overset{\overset{H}{|}}{C}}-\underset{\underset{H}{|}}{\overset{\overset{H}{|}}{C}}-\underset{\underset{H}{|}}{\overset{\overset{H}{|}}{C}}-\underset{\underset{H}{|}}{\overset{\overset{H}{|}}{C}}-H$$

(4)
$$H-C\equiv C-\underset{\underset{H}{|}}{\overset{\overset{H}{|}}{C}}-\underset{\underset{H}{|}}{\overset{\overset{H}{|}}{C}}-H$$

50 If a hydrocarbon molecule contains a triple bond, its
    IUPAC name ends in
    1 "ane"   3 "one"
    2 "ene"   4 "yne"

51 Which compound is an organic acid?
    (1) CH_3OH (3) CH_3COOH
    (2) CH_3OCH_3 (4) CH_3COOCH_3

52 Which structural formula represents the product
    formed from the reaction of Cl_2 and C_2H_4?

$$
\begin{array}{ccc}
 & H & H \\
 & | & | \\
(1)\ H- & C- & C-H \\
 & | & | \\
 & Cl & Cl
\end{array}
$$

$$
\begin{array}{ccc}
 & Cl & Cl \\
 & | & | \\
(2)\ H- & C= & C-H
\end{array}
$$

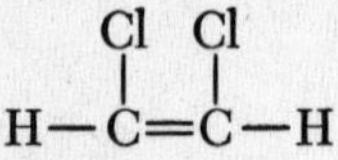

(3) $H-C \equiv C-Cl$

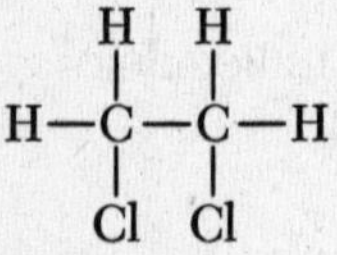

$$
\begin{array}{ccc}
 & H & H \\
 & | & | \\
(4)\ H- & C- & C-Cl \\
 & | & | \\
 & H & H
\end{array}
$$

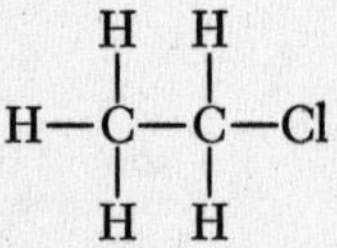

53 Which homologous series contains the compound
    toluene?
    1 alkene   3 alkyne
    2 benzene   4 alkane

Note that questions 54 through 56 have only three choices.

54 As the elements in Group 17 (VIIA) are considered in order from top to bottom, the strength of the van der Waals forces between the atoms of each successive element is
1 less
2 greater
3 the same

55 As the number of effective collisions between the reactant particles in a chemical reaction decreases, the rate of the reaction
1 decreases
2 increases
3 remains the same

56 A sealed container of nitrogen gas contains 6×10^{23} molecules at STP. As the temperature increases, the mass of the nitrogen will
1 decrease
2 increase
3 remain the same

PART 2: *This part consists of twelve groups. Choose seven of these twelve groups. Be sure to answer all questions in each group chosen. Write the answers to these questions on the separate answer sheet provided.* [35]

GROUP 1—Matter and Energy
If you choose this group, be sure to answer questions 57–61.

57 Which pair must represent atoms of the same element?
(1) $^{14}_{6}X$ and $^{14}_{7}X$ (3) $^{2}_{1}X$ and $^{4}_{2}X$
(2) $^{12}_{6}X$ and $^{13}_{6}X$ (4) $^{13}_{6}X$ and $^{14}_{7}X$

58 Which graph best represents a change of phase from a gas to a solid?

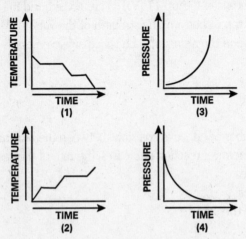

59 At 1 atmosphere and 20°C, all samples of $H_2O(\ell)$ must have the same
1 mass   3 volume
2 density   4 weight

60 The total quantity of molecules contained in 5.6 liters of a gas at STP is
(1) 1.0 mole   (3) 0.50 mole
(2) 0.75 mole   (4) 0.25 mole

61 A sample of gas has a volume of 2.0 liters at a pressure of 1.0 atmosphere. When the volume increases to 4.0 liters, at constant temperature, the pressure will be
(1) 1.0 atm   (3) 0.50 atm
(2) 2.0 atm   (4) 0.25 atm

GROUP 2—Atomic Structure

If you choose this group, be sure to answer questions 62–66.

62 Which radioactive sample would contain the greatest remaining mass of the radioactive isotope after 10 years?
   (1) 2.0 grams of ^{198}Au   (3) 4.0 grams of ^{32}P
   (2) 2.0 grams of ^{42}K   (4) 4.0 grams of ^{60}Co

63 Neutral atoms of the same element can differ in their number of
   1 neutrons   3 protons
   2 positrons   4 electrons

64 In which reaction is the first ionization energy greatest?
   (1) Na + energy → Na$^+$ + e$^-$
   (2) K + energy → K$^+$ + e$^-$
   (3) Mg + energy → Mg$^+$ + e$^-$
   (4) Al + energy → Al$^+$ + e$^-$

65 If 75.0% of the isotopes of an element have a mass of 35.0 amu and 25.0% of the isotopes have a mass of 37.0 amu, what is the atomic mass of the element?
   (1) 35.0 amu   (3) 36.0 amu
   (2) 35.5 amu   (4) 37.0 amu

66 What is the maximum number of electrons that may be present in the fourth principal energy level of an atom?
   (1) 8   (3) 18
   (2) 2   (4) 32

GROUP 3—Bonding

*If you choose this group, be sure to answer questions **67–71**.*

67 Which compound contains ionic bonds?
    (1) N_2O (3) CO
    (2) Na_2O (4) CO_2

68 What is the total number of moles of atoms in one mole of $(NH_4)_2SO_4$?
    (1) 10   (3) 14
    (2) 11   (4) 15

69 A substance was found to be a soft, nonconducting solid at room temperature. The substance is most likely
    1 a molecular solid   3 a metallic solid
    2 a network solid   4 an ionic solid

70 Two atoms with an electronegativity difference of 0.4 form a bond that is
    1 ionic, because electrons are shared
    2 ionic, because electrons are transferred
    3 covalent, because electrons are shared
    4 covalent, because electrons are transferred

71 Which species contains a coordinate covalent bond?

    (1) $\overset{\times\times}{\underset{\times\times}{\times}}\ddot{C}l\times\ddot{C}l\text{:}$ (3) $Na^+\left[\times\ddot{\ddot{C}l}\text{:}\right]^{\,-}$

    (2) $H\times\ddot{C}l\text{:}$ (4) $\left[\begin{matrix} & H & \\ H\times & \ddot{N} & \times H \\ & H & \end{matrix}\right]^{+}$

GROUP 4—Periodic Table

*If you choose this group, be sure to answer questions **72–76**.*

72 As the elements Li to F in Period 2 of the Periodic Table are considered in succession, how do the relative electronegativity and the covalent radius of each successive element compare?

  1 The relative electronegativity decreases, and the covalent radius decreases.

  2 The relative electronegativity decreases, and the covalent radius increases.

  3 The relative electronegativity increases, and the covalent radius decreases.

  4 The relative electronegativity increases, and the covalent radius increases.

73 A characteristic of most nonmetallic solids is that they are

  1 brittle

  2 ductile

  3 malleable

  4 conductors of electricity

74 In which category of elements in the Periodic Table do all of the atoms have valence electrons in the second principal energy level?

  1 Group 2 (IIA)

  2 Period 2

  3 the alkaline earth family

  4 the alkali metals family

75 Which element can form a chloride with a general formula of MCl_2 or MCl_3?

  (1) Fe   (3) Mg

  (2) Al   (4) Zn

76 Which ion has the same electron configuration as an
   H⁻ ion?
   (1) Cl⁻   (3) K⁺
   (2) F⁻   (4) Li⁺

GROUP 5—Mathematics of Chemistry

If you choose this group, be sure to answer questions **77–81.**

77 What is the molecular formula of a compound with the
   empirical formula P_2O_5 and a gram-molecular mass of
   284 grams?
   (1) P_2O_5 (3) $P_{10}O_4$
   (2) P_5O_2 (4) P_4O_{10}

78 How many molecules are in 0.25 mole of CO?
   (1) 1.5×10^{23} (3) 3.0×10^{23}
   (2) 6.0×10^{23} (4) 9.0×10^{23}

79 If the pressure and Kelvin temperature of 2.00 moles of
   an ideal gas at STP are doubled, the resulting volume
   will be
   (1) 5.60 L   (3) 22.4 L
   (2) 11.2 L   (4) 44.8 L

80 The freezing point of a 1.00-molal solution of
   $C_2H_4(OH)_2$ is closest to
   (1) +1.86°C   (3) –3.72°C
   (2) –1.86°C   (4) +3.72°C

81 The molarity (M) of a solution is equal to the

(1) $\dfrac{\text{number of grams of solute}}{\text{liter of solvent}}$

(2) $\dfrac{\text{number of grams of solute}}{\text{liter of solution}}$

(3) $\dfrac{\text{number of moles of solute}}{\text{liter of solvent}}$

(4) $\dfrac{\text{number of moles of solute}}{\text{liter of solution}}$

GROUP 6—Kinetics and Equilibrium

If you choose this group, be sure to answer questions 82–86.

82 Given the reaction at equilibrium:

$$Mg(OH)_2(s) \rightleftarrows Mg^{2+}(aq) + 2OH^-(aq)$$

The solubility product constant for this reaction is correctly written as

(1) $K_{sp} = [Mg^{2+}] [2OH^-]$
(2) $K_{sp} = [Mg^{2+}] + [2OH^-]$
(3) $K_{sp} = [Mg^{2+}] [OH^-]^2$
(4) $K_{sp} = [Mg^{2+}] + [OH^-]^2$

83 Based on Reference Table *D*, which salt solution could contain 42 grams of solute per 100 grams of water at 40°C?
1 a saturated solution of $KClO_3$
2 a saturated solution of KCl
3 an unsaturated solution of $NaCl$
4 an unsaturated solution of NH_4Cl

84 The value of the equilibrium constant of a chemical reaction will change when there is an increase in the

1 temperature
2 pressure
3 concentration of the reactants
4 concentration of the products

85 Given a saturated solution of silver chloride at constant temperature:

$$AgCl(s) \rightleftarrows Ag^+(aq) + Cl^-(aq)$$

As NaCl(s) is dissolved in the solution, the concentration of the Ag^+ ions in the solution

1 decreases, and the concentration of Cl^- ions increases
2 decreases, and the concentration of Cl^- ions remains the same
3 increases, and the concentration of Cl^- ions increases
4 increases, and the concentration of Cl^- ions remains the same

86 In which reaction will the point of equilibrium shift to the left when the pressure on the system is increased?

(1) $C(s) + O_2(g) \rightleftarrows CO_2(g)$
(2) $CaCO_3(s) \rightleftarrows CaO(s) + CO_2(g)$
(3) $2Mg(s) + O_2(g) \rightleftarrows 2MgO(s)$
(4) $2H_2(g) + O_2(g) \rightleftarrows 2H_2O(g)$

GROUP 7—Acids and Bases

If you choose this group, be sure to answer questions 87–91.

87 If an aqueous solution turns blue litmus red, which relationship exists between the hydronium ion and hydroxide ion concentrations?
(1) $[H_3O^+] > [OH^-]$
(2) $[H_3O^+] < [OH^-]$
(3) $[H_3O^+] = [OH^-] = 10^{-7}$
(4) $[H_3O^+] = [OH^-] = 10^{-14}$

88 Which metal will react with hydrochloric acid and produce $H_2(g)$?
(1) Au   (3) Mg
(2) Cu   (4) Hg

89 The concentration of hydrogen ions in a solution is 1.0×10^{-5} M at 298 K. What is the concentration of hydroxide ions in the same solution?
(1) 1.0×10^{-14} M   (3) 1.0×10^{-7} M
(2) 1.0×10^{-9} M   (4) 1.0×10^{-5} M

90 Given the reactions X and Y below:

X: $H_2O + NH_3 \rightarrow NH_4^+ + OH^-$
Y: $H_2O + HSO_4^- \rightarrow H_3O^+ + SO_4^{2-}$

Which statement describes the behavior of the H_2O in these reactions?
1 Water acts as an acid in both reactions.
2 Water acts as a base in both reactions.
3 Water acts as an acid in reaction X and as a base in reaction Y.
4 Water acts as a base in reaction X and as an acid in reaction Y.

91 In the reaction
$$H_2PO_4^- + H_2O \rightleftarrows H_3PO_4 + OH^-,$$
which pair represents an acid and its conjugate base?
(1) H_2O and $H_2PO_4^-$
(2) H_2O and H_3PO_4
(3) H_3PO_4 and OH^-
(4) H_3PO_4 and $H_2PO_4^-$

GROUP 8—Redox and Electrochemistry

If you choose this group, be sure to answer questions 92–96.

92 Based on Reference Table *N*, which half-cell has a lower reduction potential than the standard hydrogen half-cell?
(1) $Cu^{2+} + 2e^- \rightarrow Cu(s)$
(2) $Fe^{2+} + 2e^- \rightarrow Fe(s)$
(3) $I_2(s) + 2e^- \rightarrow 2I^-$
(4) $Cl_2(g) + 2e^- \rightarrow 2Cl^-$

93 Which equation represents a redox reaction?
(1) $2Na^+ + S^{2-} \rightarrow Na_2S$
(2) $H^+ + C_2H_3O_2^- \rightarrow HC_2H_3O_2$
(3) $NH_3 + H^+ + Cl^- \rightarrow NH_4^+ + Cl^-$
(4) $Cu + 2Ag^+ + 2NO_3^- \rightarrow 2Ag + Cu^{2+} + 2NO_3$

94 Which half-reaction shows both the conservation of mass and the conservation of charge?
(1) $Cl_2 + 2e^- \rightarrow 2Cl^-$
(2) $Cl_2 \rightarrow Cl^- + 2e^-$
(3) $2Br^- + 2e^- \rightarrow Br_2$
(4) $Br^- \rightarrow Br_2 + 2e^-$

95 In an electrolytic cell, to which electrode will a positive
ion migrate and undergo reduction?
1 the anode, which is negatively charged
2 the anode, which is positively charged
3 the cathode, which is negatively charged
4 the cathode, which is positively charged

96 Given the equation:

$$\_KMnO_4 + \_HCl \rightarrow \_KCl + \_MnCl_2 + \_Cl_2 + \_H_2O$$

What is the coefficient of H_2O when the equation is
correctly balanced?
(1) 8
(2) 2
(3) 16
(4) 4

GROUP 9—Organic Chemistry

If you choose this group, be sure to answer questions **97–101.**

97 Which is a product of a condensation reaction?
(1) O_2
(2) CO_2
(3) H_2
(4) H_2O

98 A molecule of ethane and a molecule of ethene both
have the same
1 empirical formula
2 molecular formula
3 number of carbon atoms
4 number of hydrogen atoms

99 Which is an example of a monohydroxy alcohol?
   1 methanal
   2 methanol
   3 glycol
   4 glycerol

100 Which property is generally characteristic of an organic compound?
   1 low melting point
   2 high melting point
   3 soluble in polar solvents
   4 insoluble in nonpolar solvents

101 Which is the structural formula of an aldehyde?

(1)
$$H-\underset{\underset{H}{|}}{\overset{\overset{H}{|}}{C}}-OH$$

(2)
$$H-C\underset{H}{\overset{O}{<}}$$

(3)
$$H-C\underset{OH}{\overset{O}{<}}$$

(4)
$$H-\underset{\underset{H}{|}}{\overset{\overset{H}{|}}{C}}-O-\underset{\underset{H}{|}}{\overset{\overset{H}{|}}{C}}-H$$

GROUP 10—Applications of Chemical Principles

If you choose this group, be sure to answer questions **102–106.**

102 Given the reaction for the nickel-cadmium battery:

$$2NiOOH + Cd + 2H_2O \rightarrow 2Ni(OH)_2 + Cd(OH)_2$$

Which species is oxidized during the discharge of the battery?

(1) Ni^{3+} (3) Cd

(2) Ni^{2+} (4) Cd^{2+}

103 Petroleum is a natural source of

1 alcohols   3 esters

2 hydrocarbons   4 ketones

104 Which acid is formed during the contact process?

(1) HNO_2 (3) H_2SO_4

(2) HNO_3 (4) H_2S

105 Group 1 and Group 2 metals are obtained commercially from their fused compounds by

1 reduction with CO

2 reduction by heat

3 reduction with Al

4 electrolytic reduction

106 Which balanced equation represents a cracking reaction?

(1) $C_4H_{10} \rightarrow C_2H_6 + C_2H_4$

(2) $C_4H_8 + 6O_2 \rightarrow 4CO_2 + 4H_2O$

(3) $C_4H_{10} + Br_2 \rightarrow C_4H_9Br + HBr$

(4) $C_4H_8 + Br_2 \rightarrow C_4H_8Br_2$

GROUP 11—Nuclear Chemistry

If you choose this group, be sure to answer questions 107–111.

107 Bombarding a nucleus with high-energy particles that change it from one element into another is called
1 a half-reaction
2 a breeder reaction
3 artificial transmutation
4 natural transmutation

108 Given the nuclear reaction:

$$^{14}_{7}N + ^{4}_{2}He \rightarrow ^{1}_{1}H + X$$

Which isotope is represented by the X when the equation is correctly balanced?
(1) $^{17}_{8}O$ (3) $^{17}_{9}F$
(2) $^{18}_{8}O$ (4) $^{18}_{9}F$

109 Which conditions are required to form $^{4}_{2}He$ during the fusion reaction in the sun?
1 high temperature and low pressure
2 high temperature and high pressure
3 low temperature and low pressure
4 low temperature and high pressure

110 The temperature levels in a nuclear reactor are maintained primarily by the use of
1 shielding   3 moderators
2 coolants   4 control rods

111 In an experiment, radioactive $Pb^*(NO_3)_2$ [* indicates radioactive Pb^{2+} ions] was added to the following equilibrium system:

$$PbCl_2(s) \rightleftarrows Pb^{2+}(aq) + 2Cl^-(aq)$$

When equilibrium was reestablished, some of the $PbCl_2(s)$ was recovered from the system and dried. Testing showed the $PbCl_2(s)$ was radioactive. Which statement is best supported by this result?

1 At equilibrium, the rates of chemical change are equal.
2 At equilibrium, the rates of chemical change are unequal.
3 The process of dynamic equilibrium is demonstrated.
4 The process of dynamic equilibrium is not demonstrated.

GROUP 12—Laboratory Activities

*If you choose this group, be sure to answer questions **112–116**.*

112 Which diagram represents an Erlenmeyer flask?

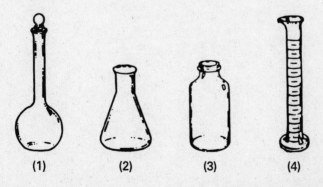

(1)   (2)   (3)   (4)

113 Salt *A* and salt *B* were dissolved separately in 100-millimeter beakers of water. The water temperatures were measured and recorded as shown in the table below.

	Salt A	Salt B
Initial water temperature:	25.1°C	25.1°C
Final water temperature:	30.2°C	20.0°C

Which statement is a correct interpretation of these data?
1 The dissolving of only salt *A* was endothermic.
2 The dissolving of only salt *B* was exothermic.
3 The dissolving of both salt *A* and salt *B* was endothermic.
4 The dissolving of salt *A* was exothermic and the dissolving of salt *B* was endothermic.

114 Which measurement has the greatest number of significant figures?
(1) 6.060 mg   (3) 606 mg
(2) 60.6 mg   (4) 60600 mg

115 The graph below was constructed by a student to show the relationship between temperature and time as heat was uniformly added to a solid below its melting point.

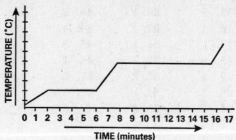

What is the total length of time that the solid phase was in equilibrium with the liquid phase?
(1) 6 min   (3) 8 min
(2) 10 min   (4) 4 min

116 The following data were collected at the end-point of a titration performed to find the molarity of an HCl solution.

Volume of acid (HCl) used = 14.4 mL
Volume of base (NaOH) used = 22.4 mL
Molarity of standard base (NaOH) = 0.20 M

What is the molarity of the acid solution?
(1) 1.6M   (3) 0.31 M
(2) 0.64 M   (4) 0.13 M

ANSWER KEY
JUNE 1994

PART 1

1. 3	15. 2	29. 2	43. 2
2. 1	16. 4	30. 4	44. 4
3. 2	17. 1	31. 2	45. 3
4. 4	18. 4	32. 3	46. 1
5. 4	19. 4	33. 2	47. 2
6. 3	20. 3	34. 1	48. 3
7. 1	21. 1	35. 2	49. 3
8. 2	22. 3	36. 4	50. 4
9. 3	23. 2	37. 3	51. 3
10. 2	24. 4	38. 1	52. 1
11. 1	25. 1	39. 2	53. 2
12. 3	26. 4	40. 4	54. 2
13. 1	27. 2	41. 1	55. 1
14. 4	28. 1	42. 3	56. 3

PART 2

57. 2	75. 1	93. 4	111. 3
58. 1	76. 4	94. 1	112. 2
59. 2	77. 4	95. 3	113. 4
60. 4	78. 1	96. 1	114. 1
61. 3	79. 4	97. 4	115. 4
62. 4	80. 2	98. 3	116. 3
63. 1	81. 4	99. 2	
64. 3	82. 3	100. 1	
65. 2	83. 4	101. 2	
66. 4	84. 1	102. 3	
67. 2	85. 1	103. 2	
68. 4	86. 2	104. 3	
69. 1	87. 1	105. 4	
70. 3	88. 3	106. 1	
71. 4	89. 2	107. 3	
72. 3	90. 3	108. 1	
73. 1	91. 4	109. 2	
74. 2	92. 2	110. 2	

ANSWERS AND EXPLANATIONS
JUNE 1994

PART 1

1. **3** By definition the word *fusion* means melting. During the process of melting, 80 calories of energy is needed to melt 1 gram of ice. The 80 calories per gram required for melting is termed the *heat of fusion*. Condensation (choice 1) occurs when gases change to liquid, crystallization (choice 2) is the process through which solids are formed when precipitation takes place in an aqueous solution, and formation (choice 4) is the process through which elements are assembled to form a solid molecule.

2. **1** Sublimation is the process by which solids become gases, or vice versa, without going through the liquid phase. Choice 1 is the correct answer because it is the only one that shows this kind of phase change (solid → gas).

3. **2** Compounds can be broken down into elements; however, elements cannot be broken down any further. Choice 2 is the correct answer because it is the only element (Al) given. Choices 1, 3, and 4 are all compounds.

    Ammonia = NH_3
    Methane = CH_4
    Methanol = CH_3OH

4. **4** Geometrically arranged particles are found in solids. Choice 4 is the correct answer because it is the only one in the solid state.

5. **4** The prefix *kilo* means 10^3 or 1,000. Therefore, 35 kilocalories is equal to $35 \times 1,000$ calories = 35,000 calories.

6. **3** The electron configuration of hydrogen (H) is $1s^1$. The electron configuration of helium (He) is $1s^2$. According to the electron configurations of H and He, hydrogen atoms have a half-filled s orbital, whereas helium atoms have a filled s orbital. The maximum number of electrons that can be placed in the s orbital is 2. Choices 1, 2, and 4 do not

agree with the electron configuration of hydrogen. Both hydrogen and helium have the same principal energy level of 1 (contradicts choice 1). Hydrogen atoms have only one atom (contradicts choice 2). Hydrogen atoms do not have a filled s shell because the maximum number of electrons that can be placed in an s shell is two (choice 4).

7. 1   Opposite charges attract each other. Therefore, positive electrodes attract negative charges. The particle $_{-1}^{0}e$ is negatively charged. Consequently, $_{-1}^{0}e$ is attracted to positive electrodes. All other choices are either positively charged or neutral in nature.

8. 2   The correct order in which electrons fill the energy levels is as follows: $1s^2 2s^2 2p^6 3s^2 3p^6 4s^2 3d^{10}$. Atoms in an excited state have an electron configuration in which electrons skip energy levels and move to a higher level. Choice 2 is the correct answer because the electron in $3s^1$ skipped the $2p^6$ and moved up to a higher level.

9. 3   Energy is released when electrons fall from a higher energy level to a lower energy level in an atom. Choice 3 is the correct choice because the electron falls from a higher energy level ($3p$) to a lower energy level ($1s$). All other choices indicate that electrons are moving from a lower energy level to a higher energy level. When electrons get excited and jump to a higher energy level, energy is absorbed.

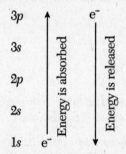

10. 2   The definition of *atomic number* is the number of protons in an atom. Every element has a different atomic number because each has a unique number of protons.

ATOM

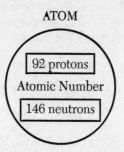

92 protons

Atomic Number

146 neutrons

MASS NUMBER = 92 + 146 = 238

11. 1 The mass of an electron is approximately $\frac{1}{1,836}$ of the mass of a proton. (The mass of a proton is 1 amu.) A proton can be also represented as $_1^1H$ because a hydrogen atom has a single proton.

12. 3 An alpha decay is described as an emission of an alpha particle from the nucleus of an atom. An alpha particle resembles a helium atom ($_2^4He$). Choice 3 is the only one that shows $_2^4He$ as a product of the reaction.

13. 1 According to the trend observed in the periodic table, the electronegativity of elements increases from left to right (see Reference Table K). Choice 1 is the only graph that shows a steady increase in the electronegative value as atomic number increases (positive slope). All other graphs show negative slopes.

14. 4 According to the equation $N_{(g)} + N_{(g)} \rightarrow N_{2(g)}$ + energy, two nitrogen atoms combine to form a diatomic nitrogen molecule. In addition, energy is a product of the reaction because it is on the right side of the equation. Choice 4 defines the events of the reaction correctly.

15. 2 Hydrogen bonding is an attraction between the hydrogen of one molecule and the fluorine, oxygen, or nitrogen of an adjacent molecule. Based on the FON (fluorine, oxygen, and nitrogen) rule, hydrogen bonds can only form between molecules that have hydrogen and at least one fluorine, oxygen, or nitrogen bonded to it. Therefore, of the choices given, a hydrogen bond can only exist between two H_2O mole-

cules because all water molecules possess both hydrogen and oxygen. All other choices have molecules that violate the FON rule. S, Se, and Te cannot form hydrogen bonds.

$$\delta^-$$

δ = partial charge

covalent bond (intramolecular)

hydrogen bond (intermolecular)

16. 4 A chemical formula is used to express the composition of elements and compounds. In the case of compounds, the formula tells us the exact ratio in which elements combine to form a compound. For example, the chemical formula for methane gas is CH_4. According to the formula, methane has one carbon and four hydrogen atoms.

17. 1 Water is a polar molecule. Due to a great difference in electronegativity between the oxygen and hydrogen atoms in water molecules, the oxygen acquires a partial negative charge and the hydrogens are positively charged. Consequently, the water molecule is bent in shape. When calcium chloride ($CaCl_2$) is dissolved in water, the Ca^{2+} ions (positively charged) are attracted to the negatively charged oxygen atoms of the water molecules.

18. 4 A Na^+ ion has one *less* electron than a Na atom. Sodium ions (Na^+) have 11 protons and 10 electrons, whereas sodium atoms (Na) have 11 protons and 11 electrons. Because the sodium ion and atom have different numbers of electrons, their electron configuration (arrangement) is also different. The Na atoms have one additional energy level to accommodate the extra electron compared to Na^+ ions. The difference in electron configuration (especially the difference in valence electrons) gives the Na atom and the Na ion different chemical properties.

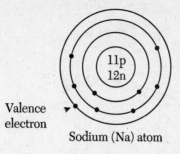

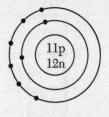

Valence electron ▶

Sodium (Na) atom

Sodium ion (Na⁺)
(Note that the ion has
lost the valence electron)

19. **4** All elements with atomic numbers between 1 and 83 have *at least* one stable isotope. However, all elements with an atomic number of 84 or higher lack any stable isotope. Of the choices given, only polonium has an atomic number that is equal to or greater than 84. In fact, polonium's atomic number is 84 (see the periodic table).

20. **3** Halogens belong to group 17 in the periodic table. Of the choices given, iodine (I) is part of group 17. Other halogens in the same group are fluorine (F), chlorine (Cl), bromine (Br), and astatine (At).

21. **1** According to Reference Table K, ionization energy increases as we move from left to right in a period. We also know that nonmetals are on the right side of the periodic table. Therefore, we can conclude that nonmetals have a high ionization energy. This makes sense because nonmetals tend to gain electrons to fill their valence shells. They do not lose electrons easily. The energy needed to remove an electron from an atom of a nonmetal (its ionization energy) is high. Nonmetals are also poor conductors of electricity. Metals are considered good conductors.

22. **3** Metalloids are found on either side of the bold line that separates metals from nonmetals in the periodic table. In group 15, As and Sb are found on either side of the bold line. Keep in mind that all the elements that sit directly above the bold line are metalloids (B, Si, As, Te, At), but of the elements below the bold line, only Ge and Sb are metalloids.

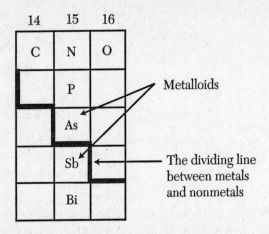

	14	15	16
	C	N	O
		P	
		As	
		Sb	
		Bi	

Metalloids

The dividing line between metals and nonmetals

23. 2 Reference Table P indicates the atomic radii of metal ions. According to Table P, potassium (K) has the largest radius.

Element	Atomic radius (Å)
Sodium (Na)	1.54
Potassium (K)	2.27
Magnesium (Mg)	1.60
Calcium (Ca)	1.97

24. 4 An element with a total of four electrons in its outermost energy level (valence shell) must have the electron configuration ns^2np^2. A quick look at the periodic table reveals that group 14 elements have the electron configuration ns^2np^2. For example, carbon's electron configuration is $1s^22s^22p^2$.

25. 1 Group 1 elements are all metals. Because they have a low first ionization energy (see Reference Table K), group 1 metals can easily give up an electron (high reactivity) and form ionic bonds with nonmetals to create ionic compounds. All ionic compounds are stable in nature.

26. 4 The formula weight of 1 mole of LiF is 26 grams per mole.

$$Li = 7 \text{ g/mol}$$
$$+ \ F = 19 \text{ g/mol}$$
$$\overline{\ 26 \text{ g/mol}}$$

Therefore, 39 grams of LiF contains the following number of moles:

Moles = given mass/formula weight per mole
= 39 g/26 g/mol = 1.5 mol

Hint: When solving for moles always divide the given value by the value of 1 mole (in the same units).

27. 2 One mole of gas at STP occupies 22.4 liters of volume. Therefore, 0.5 mole contains exactly half the volume of 22.4 liters, or 22.4 L/mol × 0.5 mol = 11.2 L. Choice 2 is the correct answer because it indicates the volume to be 11.2 L. Choices 1, 3, and 4 all represent 1 mole.

28. 1 Use the following formula to solve the problem:

$$\text{Molarity}_{(before)} \times \text{volume}_{(before)} = \text{molarity}_{(after)} \times \text{volume}_{(after)}$$
$$(12 \text{ M})(0.5 \text{ L}) = (M)(1 \text{ L})$$
$$M = 6$$

Therefore, the new molarity is 6.0M.

29. 2 According to the problem, the gram formula mass of Fe_2O_3 is 160 g/mol. The mass of the three oxygen atoms (O_3) in Fe_2O_3 is 3 × 16 g = 48 g. Therefore, the percent by mass of oxygen in Fe_2O_3 is 48 g/160 g × 100% = 30%.

30. 4 According to the balanced equation, CO_2 and O_2 react with each other in a mole ratio of 1:1 (for every six CO_2s that react, the reaction produces six O_2s). In the problem, we are given 32 grams of O_2, or 32 grams/32 grams per mole = 1 mole of oxygen. Because CO_2 and O_2 react in a 1:1 ratio, the amount of CO_2 used in the reaction must be 1 mole. At STP, 1 mole of gas occupies 22.4 liters.

31. 2 By increasing the surface area, we can have more HCl molecules come in contact with more Mg atoms. When Mg is in a solid form, the surface area is limited. However, in its powdered form, surface area is increased.

32. 3 Reference Table I gives the heat of reaction, or ΔH. If ΔH is negative, energy is released when products are formed, and the products have less energy than the reactants. On the other hand, if energy is absorbed when products are formed, ΔH is positive, and the products have more energy than the reactants. According to Table I, only the

reaction in choice 3 has a positive ΔH (+3.5 kcal), indicating that the products have a higher energy content than the reactants.

33. 2 The correct formula for free energy change in a chemical reaction is

$\Delta G = \Delta H + T\Delta S$

Free energy = heat of reaction$_{(enthalpy)}$ + temperature (in K) × change in entropy

34. 1 By definition, all catalysts speed up the rate of reaction by lowering the activation energy. Catalysts make no other changes in the characteristics of reactions.

35. 2 Gases dissolve best in water under conditions of high pressure and low temperature. Under high pressure, it is easier to squeeze gas molecules into water. According to Reference Table D, solubility of gases increases as temperature decreases. Choice 2 is the only answer that states these conditions.

36. 4 An acid is defined as a substance that produces H^+ ions, and a base as one that produces OH^- ions in solution. When H^+ ions outnumber OH^- ions in solution, we say that the solution is acidic. When H^+ and OH^- ions are equal in number in a solution, the solution is said to be neutral. When OH^- outnumber H^+ ions in solution, the solution is defined as basic.

37. 3 A Brönsted-Lowry acid is a substance that can donate a proton (H^+). In the given reaction, HCl must be the Brönsted-Lowry acid because it donates a proton to the H_2O molecule. By donating a proton, HCl becomes Cl^-, and H_2O becomes H_3O^+.

$$HCl + H_2O \rightarrow H_3O^+ + Cl^-$$

H^+ ion is transferred
from HCl to H_3O^+

38. 1 According to the law of dissociation, the higher the acid dissociation constant (K_a), the stronger the acid is and the better it acts as a conductor of electricity. Reference Table L summarizes the K_a values of several acids. Table L is organized so that strong acids (high K_a val-

ues) are listed at the top and weak acids (low K_a values) are listed at the bottom. Choice 1 (H_2S) is the correct answer because it has the lowest K_a value of the four choices given. As a result, H_2S dissociates the least. Acids that dissociate the least are the poorest conductors of electricity.

39. 2 The Arrhenius theory defines acids as substances that can release H^+ ions in water. Recall that at a pH of 7 (neutral), there is an equal concentration of H^+ and OH^- in water. Therefore, all acidic aqueous solutions must have an excess of H^+ ions.

40. 4 All basic solutions have a pH greater than 7. However, the highest basic pH possible is 14 (remember that the pH scale ranges from 0 [acidic] to 14 [basic]). The pH value that is the most basic is the one that is closest to pH 14.

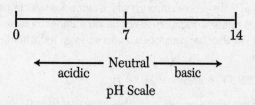

41. 1 When a solution reaches neutrality, the number of moles of acid is equal to the number of moles of base in it. Based on this fundamental rule, we can use the following formula to calculate the molarity of HNO_3:

$$\text{Volume}_{(base)} \times \text{molarity}_{(base)} = \text{volume}_{(acid)} \times \text{molarity}_{(acid)}$$
$$6.0 \text{ ml} \times 0.5 \text{ M} = 3.0 \text{ ml} \times \text{molarity}_{(acid)}$$
$$\text{Molarity}_{(acid)} = 1.0 \text{ M}$$

42. 3 According to the question, KOH (a base) is added to HCl (an acid). Therefore, the initial solution must have been acidic in nature (pH below 7). As base was added to the acidic solution, the pH of the solution rose toward neutrality.

$$HCl + KOH \rightarrow H_2O + KCl$$

This equation clearly shows that a strong acid (HCl) reacts with a strong base (KOH) to form water and a salt (KCl). This reaction is called a *neutralization reaction*. All neutralization reactions of a strong acid and a strong base lead to a pH of 7.

43. 2 Oxygen has an oxidation number of –2. In rare instances, oxygen can also have an oxidation number of –1 (e.g., in peroxide [H_2O_2]) and +2 (in HF). Because –1 and +2 are not answer choices, the only plausible answer is choice 2.

44. 4 Reduction is the process through which atoms gain electrons. When atoms gain electrons, the oxidation number decreases. All of the choices show a gain of electrons; they are written on the right side of the equation. However, only one of them correctly shows a decrease in the oxidation number following addition of electrons to an atom. The correct choice is 4, in which the oxidation number of Br_2 changes from 0 to –1.

45. 3 Oxidation is the process through which atoms lose electrons. When atoms go through oxidation, the oxidation number increases (the opposite of what happens when atoms undergo reduction). In the reaction

$$2Na + 2H_2O \rightarrow 2Na^+ + 2OH^- + H_2,$$

the sodium atoms' oxidation number increased from 0 (Na as an element has no charge) to +1 (in the Na^+ ion state). Therefore, the substance that oxidized is Na.

46. 1 By definition, oxidation is the loss of electrons. (Remember the mnemonic LEO and GER: loss of electrons, oxidation; gain of electrons, reduction.) Only choice 1 indicates a loss of electrons.

47. 2 Reference Table N, also called the *reduction potential table*, lists atoms based on their tendencies to be reduced. A reducing agent is an atom that forces another atom to gain electrons and in the process oxidizes itself (loses electrons). According to Table N, the atoms that tend to gain electrons (reduction) are indexed toward the top of the list. On the other hand, the atoms that tend to lose electrons (oxidation) are listed at the bottom of Table N. The strongest reducing agents are found near the bottom of Table N. Of the four choices given, Sr (strontium) is listed closest to the bottom of Table N. Therefore, Sr is the strongest reducing agent.

48. 3 An electrochemical cell or battery requires that electrons flow through a wire to generate electricity. Flowing electrons can be generated only if a spontaneous redox reaction is carried out. Two chemicals are present inside batteries: One performs oxidation (gives up its electrons), whereas the other chemical accepts the electrons to complete reduction. The simultaneous occurrence of oxidation and reduction reactions is called a *redox reaction.*

49. 3 Butane is a saturated molecule in which only single bonds are present. In the four choices given, the only molecule that has no bonds other than single bonds is the structure represented in choice 3. All other choices have either double or triple bonds in addition to single bonds as part of their structure.

Suffix	Bonding characteristic
-ane	All single bonds
-ene	At least one double bond
-yne	At least one triple bond

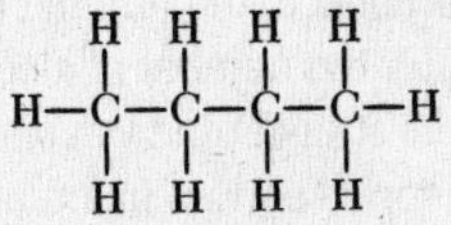

Butane (C_4H_{10})

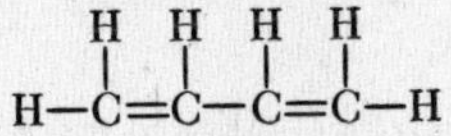

1,3 Butadiene (C_4H_8)

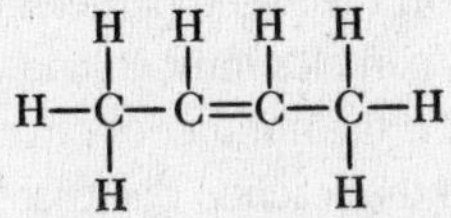

2-Butene (C_4H_8)

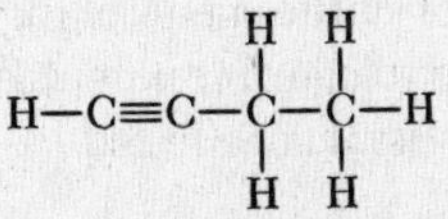

1-Butyne (C_4H_6)

50. 4 All compounds that contain a triple bond have IUPAC names that end with the suffix *-yne.*

51. 3 All organic acids have a carboxylic group (—COOH) as part of their structures. Choice 3 is the only one that correctly represents an organic acid.

Functional group	Type of molecule
—COOH	Carboxylic acid
—COH	Aldehyde
—CO—	Ketone
—COO—	Ester
—OH	Alcohol

52. 1 Chlorine (Cl_2) and ethene (C_2H_4) react in an addition reaction to form 1, 2-dichloroethane.

53. 2 Toluene is a six-carbon molecule that has a ring structure. The only other molecule that has a ring structure is benzene. The word *homologous* means similar. Therefore, the correct answer is choice 2 because toluene and benzene are similar in structure. They are both members of the benzene family.

54. 2 van der Waals forces are intermolecular forces. The strength of van der Waals forces is directly proportional to the number of electrons in an atom. In other words, the greater the number of electrons in atoms (or the larger the atoms), the stronger the van der Waals forces. As elements are considered from top to bottom in group 17, the atoms are successively larger. Therefore, the van der Waals forces are greater (choice 2).

55. 1 Particles must come in contact with each other to react. The contact allows for bonds to break and reform between particles to produce new compounds. In other words, particles must collide with each other for a reaction to take place. If the number of effective collisions between particles decreases, then the reaction decreases as well.

56. 3 The temperature is the average of kinetic energy of gas molecules. When temperature increases, the volume, pressure, and kinetic energy also increase. However, mass is not affected.

PART 2

Group 1—Matter and Energy

57. 2 All atoms of the same element must have an identical atomic number (the number of protons in the nucleus). The atomic number is repre-

sented in the periodic table by the subscript number to the left of the element's symbol. Of the choices given, only the pair of elements in choice 2 represent the same element. Both members of the pair have an atomic number of 6.

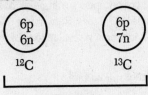

Isotopes

Note that the difference between isotopes is the number of neutrons each isotope possesses.

58. 1 When water boils, water molecules are converted from liquid phase to gaseous phase. From our daily life experience, we know that boiling, or going from liquid to gas phase, requires an increase in temperature. Therefore, if we want to proceed in the reverse direction, we must lower the temperature. For example, if we want water (liquid phase) to freeze (solid state), we lower the temperature. Based on this observation, we can conclude that temperature must be lowered to change a gas to a solid.

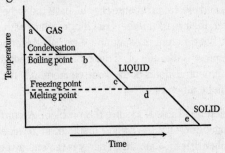

In sections a, c, and e the temperature decreases. Therefore, there is a decrease in the kinetic energy as well.

In sections b and d both temperature and kinetic energy remain the same.

59. 2   All samples of water at 1 atm and 20°C have the same density. The mass, weight, and volume of water changes depending on the type of container which holds it.

60. 4   At STP, 1 mole of gas occupies a volume of 22.4 liters. Therefore, 5.6 liters of gas contains 0.25 moles. Remember this general rule: To solve for moles, divide the given value (5.6 L) by the volume of 1 mole (22.4 L).

$5.6 \text{ L}/22.4 \text{ L} \times 1 \text{ mol} = 0.25 \text{ mol}$

61. 3   Use Boyle's law to solve this problem.

$P_{(\text{before})} \times V_{(\text{before})} = P_{(\text{after})} \times V_{(\text{after})}$, where P = pressure and V = volume

$1 \text{ atm} \times 2 \text{ L} = P_{(\text{after})} \times 4 \text{ L}$

$P_{(\text{after})} = 0.50 \text{ atm}$

Group 2—Atomic Structure

62. 4   You must use Reference Table H to solve this problem. Table H tells us the half-lives of the isotopes in the answer choices. The isotope with the longest half-life has the greatest remaining mass after 10 years. In this case, the starting mass of each isotope is irrelevant to the question. Based on Table H, ^{60}Co has the longest half-life.

63. 1   Neutral atoms of the same element must have the same number of protons and electrons because they all have the same atomic number. However, their mass numbers can differ (atoms with the same atomic number but different mass numbers are called *isotopes*). Because mass number is the sum of all the protons and neutrons in the atomic nuclei, in neutral atoms of the same element only the number of neutrons can be different.

64. 3   Use Reference Table K to solve this problem. According to Table K, Mg (magnesium) has the highest first ionization value of 176 kcal/mol. Remember that the ionization energy is the energy needed to strip the first electron from any given element (as indicated in the answer choices).

65. 2   A mass number is always a whole number. We can find the atomic mass by solving for the weighted average of all the isotopes of an ele-

ment. In this question, only two isotopes are stated, 35.0 amu (atomic mass units) and 37.0 amu isotopes.

$$0.75 \ (75\%) \times 35.0 \text{ amu} = 26.25$$
$$+ \ \ 0.25 \ (25\%) \times 37.0 \text{ amu} = 9.25$$

Weighted average = 35.5 amu

66. 4 The maximum number of electrons in any energy shell can be found by using the following formula:

Maximum number of electrons = $2n^2$, where n = the number of the energy level

Therefore, the maximum number of electrons in the fourth energy level is $2(4)^2 = 32$.

Group 3—Bonding

67. 2 Ionic bonds are formed between a metal and a nonmetal. Na_2O is the only answer choice that is composed from a metal and a nonmetal. All other answer choices have covalent bonds because they are made of two nonmetals. Also, in an ionic bond, the difference in electronegativity between the elements should be 1.7 or greater. Reference Table K shows that only Na_2O has a difference greater than 1.7 (it is 2.6).

68. 4 There are 15 moles of atoms. Simply add up all the atoms in the molecule $(NH_4)_2SO_4$.

N atoms = $1 \times 2 = 2$
H atoms = $4 \times 2 = 8$
S atoms $\quad = 1$
O atoms $\quad = 4$
Total atoms $\quad = 15$

69. 1 The substance described in the question is a molecular solid. Molecular solids are held together by weak van der Waals, or dipole-dipole, forces in haphazard arrangement. They are characteristically soft (if solid), with lower melting points than network solids, and are nonconductors.

70. 3 An electronegative difference of less than 1.7 between two elements means that a covalent bond forms between them. In covalent bonds,

electrons are shared between atoms. Ionic bonds are formed between atoms whose electronegative difference is greater than 1.7. In ionic bonds, electrons are transferred from metal ions to nonmetal ions.

71   4   A coordinate covalent bond and a regular covalent bond are similar in the sense that electrons are shared. In a regular covalent bond, both atoms contribute an electron. In a coordinate covalent bond, however, only one of the atoms contributes both electrons. The correct answer is choice 4 because nitrogen supplies both electrons to the covalent bond that it forms with the hydrogen when NH_4^+ is formed.

Group 4—Periodic Table

72.   3   As we move from left (Li) to right (F) across period 2 in the periodic table, the electronegativity increases and the covalent radius decreases. See Reference Tables K and P.

73.   1   Of the four answer choices given, only choice 1 is a characteristic of a nonmetallic solid. All the other choices are characteristics of metallic solids.

74.   2   According to the periodic table, elements in period 2 have valence electrons in the second energy level. Choices 1, 3, and 4 are examples of groups. As we move down the group, the principal energy levels increase. Therefore, it is impossible to have all the valence electrons in the second energy level within a group.

75.   1   Chlorine (Cl) has a charge of –1. In MCl_2 and MCl_3, the charge of M must be +2 and +3, respectively. The correct answer is choice 1 because only Fe can have a charge of either +2 or +3.

76.   4   An H^- ion has an electron configuration of $1s^2$. An ion with a similar electron configuration is Li^+. The normal electron configuration of Li is $1s^2 2s^1$. When a Li^+ ion is formed, an electron is lost, leading to a configuration of $1s^2$.

Group 5—Mathematics of Chemistry

77. **4** Choices 2 and 3 can be ruled out as incorrect answers because their molecular formulas are not in the ratio of 2:5 (the empirical formula is P_2O_5). The molecular mass of P_2O_5 is 142 g.

$P = 2 \times 31 = 62$
$O = 5 \times 16 = 80$
$P_2O_5 = 142$ g

The mass of the molecule in question is 284, or twice the mass in the empirical formula. Therefore, the molecular formula must be twice as large as the empirical formula, or P_4O_{10} (choice 4).

78. **1** One mole of CO at STP contains 6×10^{23} molecules. Therefore, in 0.25 mole (¼ of 1 mole), there are $0.25(6.0 \times 10^{23}$ molecules$) = 1.5 \times 10^{23}$ molecules.

79. **4** Boyle's law states that pressure and volume of gases are indirectly related to each other. Charles' law states that the temperature and volume are directly related to each other. In the given problem, both pressure and temperature are doubled. To solve for the new pressure and temperature you must use the combined gas law formula:

$$\frac{P_1V_1}{T_1} = \frac{P_2V_2}{T_2}$$

$P_1 = 1$ atm (STP)
$P_2 = 2$ atm (pressure at STP doubled)
$T_1 = 273$ K (STP)
$T_2 = 546$ K (temperature at STP doubled)
$V_1 = 44.8$ L (1 mole = 22.4 L of gas at STP)
$V_2 = ?$

Using the values above, solve for V_2.

$$\frac{1 \text{ atm} \times 44.8 \text{ L}}{273 \text{ K}} = \frac{2 \text{ atm} \times V_2}{576 \text{ K}}$$
$$V_2 = 44.8 \text{ L}$$

80. 2 According to Reference Table A, the molal freezing point of water is 1.86°C. In other words, a 1-molal solution induces water to freeze at –1.86°C. Because $C_2H_4(OH)_2$ is a dihydroxy alcohol, it does not dissociate in water. As a result, a 1-molal solution of $C_2H_4(OH)_2$ freezes at –1.86°C as well.

81. 4 The definition of molarity (M) is

$$\frac{\text{Number of moles of solute}}{\text{Liters of solution}}$$

Group 6—Kinetics and Equilibrium

82. 3 $K_{sp} = [\text{ion } X]^{\text{coefficient}} \times [\text{ion } Y]^{\text{coefficient}}$

$K_{sp} = [Mg^{2+}] \times [OH^-]^2$

83. 4 According to Reference Table D, the solubility of each of the solutes given in the four answer choices in 100 g of water is as follows:

$KClO_3$ = 15 g per 100 g of water
KCl = 40 g per 100 g of water
NaCl = 38 g per 100 g of water
NH_4Cl = 46 g per 100 g of water

If 42 g of solute is dissolved in 100 g of water at 40°C, only one of the four solutions remains unsaturated. That solution is NH_4Cl (choice 4). All other choices are supersaturated before 42 g of solute can be *completely* dissolved.

84. 1 Only temperature can change the equilibrium constant (K_{eq}) of a reaction. Changes in pressure and concentrations of reactants and products shift the direction of the reaction (Le Châtelier's principle), but do not affect the equilibrium constant.

85. 1 When NaCl is dissolved in AgCl solution, the concentration of Cl⁻ rises. According to Le Châtelier's principle, a reaction shifts away from (or removes) what is being added. Because Cl⁻ ions are being added, the equilibrium shifts to the left. A shift to the left decreases the concentration of Ag⁺ ions.

$$\uparrow AgCl \rightleftharpoons \downarrow Ag^+ + Cl^- \uparrow$$

Concentration   Concentration   NaCl increases
of   of   the concentration
AgCl $[Ag^+]$ ion   of $[Cl^-]$
increases   decreases

86. 2 Pressure only affects gases. When the pressure is increased, a reaction shifts from the side that has the higher number of moles of gases to the side with the lower number of moles of gases. Conversely, when the pressure is decreased, reaction shifts from the side with the lower number of moles of gases to the side with the higher number of moles. Choice 2 is the correct answer because an increase in pressure shifts the reaction to the left. Note that there is a mole of gas (CO_2) on the right side of the reaction but no moles of gas on the left side.

Group 7—Acids and Bases

87. 1 An aqueous solution turns a blue litmus red if the solution is acidic. In an acidic solution, the hydronium (H_3O^+) concentration is greater than the hydroxide (OH^-) concentration.

88. 3 When HCl reacts with active metals, hydrogen (H_2) gas is released. Of the four choices given, only magnesium (Mg) is an active element. Active elements are found in groups 1 and 2 in the periodic table. Any metal below H_2 on Reference Table N reacts with an acid and produces H_2 gas.

89. 2 According to the ionization constant of water, the concentration of H^+ and OH^- ions must always multiply to equal 1.0×10^{14}. Therefore, the hydroxide concentration (OH^-) is

$$[H^+][OH^-] = 1.0 \times 10^{-14}$$
$$[1.0 \times 10^{-5}][OH^-] = 1.0 \times 10^{-14}$$
$$[OH^-] = 1.0 \times 10^{-9}$$

90. 3 In reaction X, the water donates an H^+ ion to the NH_3 and, thus, acts as an acid. In reaction Y, the water molecule accepts a proton from HSO_4^-, thus acting as a base. In other words, water molecules are amphoteric in nature. They can act as both an acid and a base.

$$H_2O + NH_3 \longrightarrow NH_4^+ + OH^-$$

H^+ transferred

H_2O acts as an acid.

$$H_2O + HSO_4^- \longrightarrow H_3O^+ + SO_4^{2-}$$

H^+ transferred

H_2O acts as a base.

91. **4** In the given reaction

$$H_2PO_4^- + H_2O \rightleftarrows H_3PO_4 + OH^-,$$

$H_2PO_4^-$ (base) and H_3PO_4 (acid) are a conjugate acid-base pair. (The members of a conjugate pair differ only by a hydrogen atom and a charge.) The only answer that correctly matches the above acid-base conjugate pair is choice 4.

Group 8—Redox and Electrochemistry

92. **2** According to Reference Table N, the standard hydrogen half-cell has a reduction potential of zero volts.

$$2H^+ + 2e^- \rightarrow H_2 E^0 = 0.00 \text{ volt}$$

The only half-cell that has a lower reduction potential than the standard hydrogen half-cell is $Fe^{+2} + 2e^- \rightarrow Fe\ E^0 = -0.45$ volt. All other choices have reduction potential greater than zero.

93. **4** In a redox reaction, both reduction and oxidation must occur simultaneously. As a result, oxidation numbers of atoms change. In choice 4, the oxidation numbers for Cu and Ag change (Cu is oxidized whereas Ag is reduced).

$$\begin{array}{ccccccc} 0 & +1 & +7\,-2 & 0 & +2 & +7\,-2 \\ \end{array}$$
$$Cu + 2Ag^+ + 2NO_3^- \rightarrow 2Ag + Cu^{+2} + 2NO_3^-$$
$$Cu \rightarrow Cu^{+2} + 2e^- \text{ (oxidation)}$$
$$2\,Ag^+ + 2e^- \rightarrow 2Ag \text{ (reduction)}$$

Hint: In a question of this type, look for an element that exists in the free state on one side of the equation and is combined in a compound on the other side. Such a reaction must be undergoing redox.

94. 1 Conservation of charge means that there must be the same number of charges on either side of the reaction. Conservation of mass means that there are the same number of atoms of each element on either side of the reaction. Only choice 1 meets both the laws of conservation of mass and charge.

$$Cl_2 + 2e^- \rightarrow 2Cl^-$$

	$Cl_2 + 2e^-$	$2Cl^-$
Total charges:	-2	-2
Total mass:	$2Cl$	$2Cl$

95. 3 An electrolytic cell is composed of a cathode and an anode. The cathode is negatively charged, and the anode is positively charged. Therefore, a positive ion migrates to the cathode and undergoes reduction.

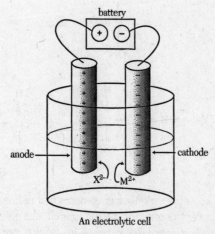

An electrolytic cell

96. 1 In a balanced equation, all the atoms of each element on both sides of the equation must be equal. The quickest way to balance any given equation is by using the trial and error method.

$$2KMnO_4 + 16HCl \rightarrow 2KCl + 2MnCl_2 + 5Cl_2 + 8H_2O$$

Based on this balanced equation, the coefficient of H_2O is 8 (answer choice 1).

Group 9—Organic Chemistry

97. 4 Condensation is the process through which water is removed during the formation of organic molecules.

98. 3 Ethane is an alkane, and its molecular formula is C_2H_6. On the other hand, ethene is an alkene, and its molecular formula is C_2H_4. Based on their formulas, the only similarity between the molecules is the number of carbons they possess (two).

99. 2 The functional group —OH represents alcohols. The word *mono* means one. Therefore, a monohydroxy alcohol contains only one —OH group. Methanol is the only answer choice that contains one —OH group.

Methanol

Acetone   Glycol   Glycerol

100. 1 You must know the various characteristics that define organic molecules to answer this question. Of the four choices given, only choice 1 is true regarding organic compounds. In addition, organic compounds are generally nonpolar, insoluble in polar solvents, and nonelectrolytes. In addition, they react more slowly than inorganic compounds.

101. 2 The functional group aldehyde is defined as —CHO. The answer that correctly depicts this functional group is choice 2.

Functional group	Type of compound
—OH	alcohol
$\overset{\overset{\displaystyle O}{\|\|}}{CH}$	aldehyde
$\overset{\overset{\displaystyle O}{\|\|}}{COH}$	carboxylic acid
—O—	ether

Group 10—Applications of Chemical Principles

102. **3** When a nickel-cadmium battery is discharged, a redox reaction occurs inside the battery.

$Ni^{+3} + e^- \rightarrow Ni^{+2}$ (reduction)
$Cd^0 \rightarrow Cd^{+2} + 2e^-$ (oxidation)

Based on the half-reactions, Cd undergoes oxidation.

103. **2** Petroleum is primarily composed of hydrocarbon molecules. These hydrocarbon molecules can be as simple as methane or as complex as a molecule that consists of more than 50 carbons. Common petroleum products include gasoline, heating oil, and tars.

104. **3** The contact process is used to commercially synthesize sulfuric acid (H_2SO_4).

$SO_3 + H_2O \rightarrow H_2SO_4$

105. **4** Elements in groups 1 and 2 are known as *alkali* and *alkaline earth metals*. These metals are extremely active. When these metals react, they form stable compounds. As a result, it is very difficult to obtain them from their fused compounds by chemical reactions. When the compounds are heated or melted, the ions move about freely. When electrodes are placed in the melted compound, the anions (negatively charged ions) and cations (positively charged ions) migrate toward electrodes of opposite charge. Group 1 and 2 ions (positively charged) combine with the negatively charged electrode and are reduced to the free element. This process is termed *electrolytic reduction*.

106. 1 Cracking refers to a type of reaction in which a larger molecule is broken into shorter molecules. The only reaction that depicts cracking is choice 1.

$$C_4H_{10} \rightarrow C_2H_6 + C_2H_4$$

All other reactions are examples of synthesis or substitution reactions.

Group 11—Nuclear Chemistry

107. 3 The key word in the answer choices is *artificial*. Changes that occur in the nucleus and cause atoms to change from one element to another are called *transmutations*.

108. 1 In a nuclear reaction, the laws of conservation of mass and charges must be followed. In other words, the sum of masses and charges on either side of the equation must equal each other. Because there is a total mass of 18 (14 from N and 4 from He) and a total charge of 9 (7 from N and 2 from He) on the left side of the equation, the right side must be the same. The only answer choice that satisfies the requirements of the equation is $^{17}_{8}O$.

109. 2 During a fusion reaction, nuclei of small atoms combine to form a larger nucleus. ^4_2He nuclei are positively charged, and it is difficult to force nuclei with like charges to combine with each other. For two positively charged nuclei to combine, they must have lots of energy. An increase in temperature raises the kinetic energy of the atoms. The atoms also combine more easily if they are "squeezed" together by an increase in pressure.

110. 2 The word *coolant* means a substance that has the ability to cool something. In a nuclear reactor, a coolant is used to maintain temperature. Shielding, moderators, and control rods are parts of a nuclear reactor.

111. 3 This is an example of a dynamic equilibrium. According to chemical laws, equilibrium is not a static state. In other words, at equilibrium the forward and reverse reactions are occurring at the same rate. In the question, it is stated that some of the $PbCl_2(s)$ became radioactive once radioactive Pb° ions were added. Radioactive $PbCl_2(s)$ can only form if the reverse reaction occurred.

Group 12—Laboratory Activities

112. **2** The Erlenmeyer flask is shown in choice 2. Choice 1 is a volumetric flask, choice 3 is a collecting bottle, and choice 4 is a graduated cylinder.

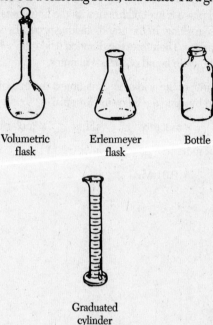

Volumetric   Erlenmeyer   Bottle
flask   flask

Graduated
cylinder

113. **4** When salt A was added, the final temperature of the water (30.2°C) was greater than the initial temperature (25.1°C). Temperature can only go up if energy or heat is added to water. Exothermic reactions release heat following the completion of a reaction. In this case, the dissolving of salt A must have been exothermic. The heat released from the exothermic reaction was used by water to increase its temperature.

When salt B was added, the final temperature of the water (20.0°C) was less than the initial temperature (25.1°C). Temperature can go down only if energy or heat is removed from water. Endothermic reactions absorb heat following the completion of a reaction. In this case, the dissolving of salt B must have been endothermic. The heat absorbed by the endothermic reaction was taken from the water, lowering the water temperature.

114. 1 Choice 1 has four significant figures (6.060), choice 2 has three significant figures (60.6), choice 3 has three significant figures (606), and choice 4 has three significant figures (60600).

115. 4 The solid phase is in equilibrium with the liquid phase during the process of melting. In the graph, melting occurs between 2 minutes and 6 minutes. Therefore, the duration when the solid phase is in equilibrium with liquid phase is 4 minutes.

116. 3 The molarity of the acid of any solution at the end of titration can be calculated by using the following formula:

$$\text{Volume}_{(acid)} \times \text{molarity}_{(acid)} = \text{volume}_{(base)} \times \text{molarity}_{(base)}$$

$$14.4 \text{ ml} \times \text{molarity}_{(acid)} = 22.4 \text{ ml} \times 0.2 \text{ M}$$

$$\text{Molarity}_{(acid)} = 0.31 \text{ M}$$

EXAMINATION
JUNE 1995

PART 1: *Answer all 56 questions in this part.* [65]

DIRECTIONS **(1–56):** *For each statement or question, select the word or expression that, of those given, best completes the statement or answers the question. Record your answer on the separate answer sheet provided.*

1 Which statement best describes the production of a chlorine molecule according to the reaction $Cl + Cl \rightarrow Cl_2 + 58$ kcal?

  1 A bond is broken, and the reaction is exothermic.

  2 A bond is broken, and the reaction is endothermic.

  3 A bond is formed, and the reaction is exothermic.

  4 A bond is formed, and the reaction is endothermic.

2 Which set of properties does a substance such as $CO_2(g)$ have?

  1 definite shape and definite volume

  2 definite shape but no definite volume

  3 no definite shape but definite volume

  4 no definite shape and no definite volume

3 Which quantity of heat does a kilocalorie represent?

  (1) 100 calories

  (2) 1000 calories

  (3) $\frac{1}{100}$ of a calorie

  (4) $\frac{1}{1000}$ of a calorie

4 Which substance can *not* be decomposed by a chemical change?

1 ammonia

2 carbon

3 methane

4 water

5 When sugar is dissolved in water, the resulting solution is classified as a

1 homogeneous mixture

2 heterogeneous mixture

3 homogeneous compound

4 heterogeneous compound

6 Which is the electron configuration of an atom in the excited state?

(1) $1s^2 2s^2 2p^2$

(2) $1s^2 2s^2 2p^1$

(3) $1s^2 2s^2 2p^5 3s^2$

(4) $1s^2 2s^2 2p^6 3s^1$

7 Which is the correct orbital notation of a lithium atom in the ground state?

(1) 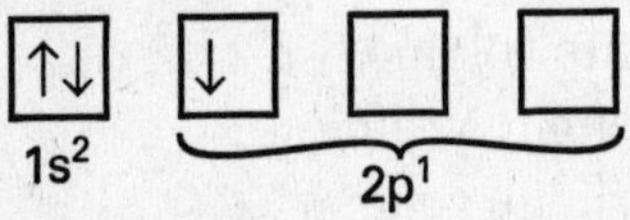
$\boxed{\uparrow\downarrow}$ $\boxed{\downarrow}$ $\boxed{}$ $\boxed{}$
$1s^2$ $2p^1$

(2)
$\boxed{\uparrow}$ $\boxed{\uparrow\downarrow}$
$1s^1$ $2s^2$

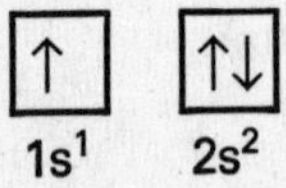

(3) 
$\boxed{\uparrow\downarrow}$ $\boxed{\uparrow\downarrow}$ $\boxed{\uparrow}$ $\boxed{\uparrow}$ $\boxed{\uparrow}$
$1s^2$ $2s^2$ $2p^3$

(4) 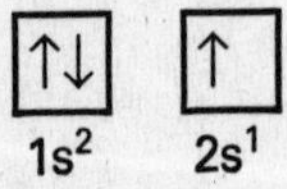
$\boxed{\uparrow\downarrow}$ $\boxed{\uparrow}$
$1s^2$ $2s^1$

8 A particle of matter contains 6 protons, 7 neutrons, and 6 electrons. This particle must be a
   1 neutral carbon atom
   2 neutral nitrogen atom
   3 positively charged carbon ion
   4 positively charged nitrogen ion

9 Which kind of particle, when passed through an electric field, would be attracted to the negative electrode?
   1 an alpha particle   3 a neutron
   2 a beta particle   4 an electron

10 What is the approximate mass of an electron?
   (1) 1 atomic mass unit   (3) $\frac{1}{1836}$ of a proton
   (2) $\frac{1}{12}$ of a C-12 atom   (4) $\frac{1835}{1836}$ of a proton

11 The mass number of an atom is always equal to the total number of its
   1 electrons, only
   2 protons, only
   3 electrons plus protons
   4 protons plus neutrons

12 Which substance is a conductor of electricity?
   (1) $NaCl(s)$ (3) $C_6H_{12}O_6(s)$
   (2) $NaCl(\ell)$ (4) $C_6H_{12}O_6(\ell)$

13 Which formula is an empirical formula?
   (1) K_2O (3) C_2H_6
   (2) H_2O_2 (4) C_6H_6

14 Molecule-ion attractions are found in
   (1) K(s)   (3) KCl(ℓ)
   (2) Kr(g)   (4) KCl(aq)

15 Which formula is correctly paired with its name?
   (1) $MgCl_2$ — magnesium chlorine
   (2) K_2O — phosphorus dioxide
   (3) $CuCl_2$ — copper (II) chloride
   (4) FeO — iron (III) oxide

16 How many moles of hydrogen atoms are present in one
   mole of $C_2H_4(OH)_2$?
   (1) 6   (3) 8
   (2) 2   (4) 4

17 In the diagram of an ammonium ion below, why is bond
   A considered to be coordinate covalent?

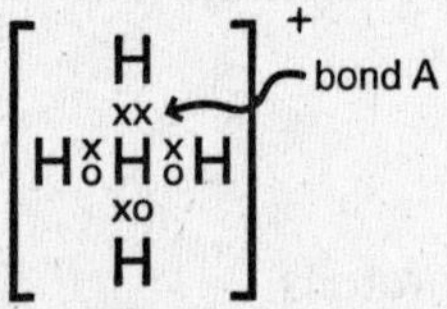

$$\left[\begin{array}{c} H \\ \overset{xx}{\longleftarrow} \\ H\overset{x}{\circ}H\overset{x}{\circ}H \\ xo \\ H \end{array} \right]^{+} \quad \text{bond A}$$

   1 Hydrogen provides a pair of electrons to be shared
     with nitrogen.
   2 Nitrogen provides a pair of electrons to be shared
     with hydrogen.
   3 Hydrogen transfers a pair of electrons to nitrogen.
   4 Nitrogen transfers a pair of electrons to hydrogen.

18 Which structural formula represents a polar molecule?

(1) $N \equiv N$

(3)
$$H - \underset{\underset{H}{|}}{\overset{\overset{H}{|}}{C}} - H$$

(2) $S = C = S$

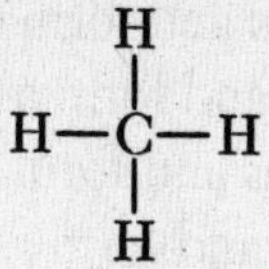

(4) 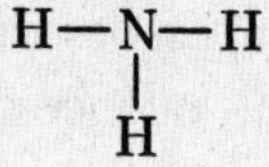
$$H - \underset{\underset{H}{|}}{N} - H$$

19 Which element is classified as a semimetal (metalloid)?
(1) Sn   (3) Pb
(2) Sb   (4) P

20 Which element in Group 15 would most likely have luster and good electrical conductivity?
(1) N   (3) Bi
(2) P   (4) As

21 Which is the electron configuration of an atom of a Period 3 element?
(1) $1s^2 2s^1$ (3) $1s^2 2s^2 2p^3$
(2) $1s^2 2s^2 2p^1$ (4) $1s^2 2s^2 2p^6 3s^1$

22 Which of the following elements has the largest covalent radius?
1 beryllium   3 calcium
2 magnesium   4 strontium

23 When a metal reacts with a nonmetal, the metal will
  1 lose electrons and form a positive ion
  2 lose protons and form a positive ion
  3 gain electrons and form a negative ion
  4 gain protons and form a negative ion

24 Which statement best describes the alkaline earth elements?
  1 They have one valence electron, and they form ions with a 1+ charge.
  2 They have one valence electron, and they form ions with a 1– charge.
  3 They have two valence electrons, and they form ions with a 2+ charge.
  4 They have two valence electrons, and they form ions with a 2– charge.

25 Which compound has the *least* ionic character?
  (1) KCl   (3) $AlCl_3$
  (2) $CaCl_2$ (4) CCl_4

26 Which statement best explains why Na is *not* found in nature?
  1 Na is very reactive, and it forms stable compounds.
  2 Na is very reactive, and it forms unstable compounds.
  3 Na is very unreactive, and it forms stable compounds.
  4 Na is very unreactive, and it forms unstable compounds.

27 What is the gram molecular mass of calcium nitrate, $Ca(NO_3)_2$?
(1) 164 g   (3) 102 g
(2) 150. g   (4) 70.0 g

28 What is the molarity of a KF(aq) solution containing 116 grams of KF in 1.00 liter of solution?
(1) 1.00 M   (3) 3.00 M
(2) 2.00 M   (4) 4.00 M

29 What is the total number of atoms in 1 mole of calcium?
(1) 1   (3) 6×10^{23}
(2) 20   (4) $20(6 \times 10^{23})$

30 What is the percent by mass of water in the hydrate $Na_2CO_3 \cdot 10H_2O$ (formula mass = 286)?
(1) 6.89%   (3) 26.1%
(2) 14.5%   (4) 62.9%

31 At STP, 32.0 liters of O_2 contains the same number of molecules as
(1) 22.4 liters of Ar   (3) 32.0 liters of H_2
(2) 28.0 liters of N_2 (4) 44.8 liters of He

32 According to Reference Table *D*, which compound's solubility decreases rapidly as the temperature changes from 10°C to 70°C?

(1) NH_4Cl (3) HCl

(2) NH_3 (4) KCl

33 Under which conditions will the rate of a chemical reaction always decrease?

1 The concentration of the reactants decreases, and the temperature decreases.

2 The concentration of the reactants decreases, and the temperature increases.

3 The concentration of the reactants increases, and the temperature decreases.

4 The concentration of the reactants increases, and the temperature increases.

34 Which is a property of a reaction that has reached equilibrium?

1 The amount of products is greater than the amount of reactants.

2 The amount of products is equal to the amount of reactants.

3 The rate of the forward reaction is greater than the rate of the reverse reaction.

4 The rate of the forward reaction is equal to the rate of the reverse reaction.

35 Based on Reference Table *G*, which reaction occurs spontaneously?

(1) $2C(s) + 3H_2(g) \rightarrow C_2H_6(g)$

(2) $2C(s) + 2H_2(g) \rightarrow C_2H_4(g)$

(3) $N_2(g) + 2O_2(g) \rightarrow 2NO_2(g)$

(4) $N_2(g) + O_2(g) \rightarrow 2NO(g)$

36 Which procedure will increase the solubility of KCl in water?
1 stirring the solute and solvent mixture
2 increasing the surface area of the solute
3 raising the temperature of the solvent
4 increasing the pressure on the surface of the solvent

37 According to the Arrhenius theory, a substance that is classified as an acid will always yield
(1) $H^+(aq)$ (3) $OH^-(aq)$
(2) $NH_4^+(aq)$ (4) $CO_3^{2-}(aq)$

38 What is the net ionic equation for a neutralization reaction?
(1) $H^+ + H_2O \rightarrow H_3O^+$
(2) $H^+ + NH_3 \rightarrow NH_4^+$
(3) $2H^+ + 2O^{2-} \rightarrow 2OH^-$
(4) $H^+ + OH^- \rightarrow H_2O$

39 Given the reaction:

$$HNO_2(aq) \rightleftarrows H^+(aq) + NO_2^-(aq)$$

The ionization constant, K_a, is equal to

(1) $\dfrac{[HNO_2]}{[H^+][NO_2^-]}$ (3) $\dfrac{[NO_2^-]}{[H^+][HNO_2]}$

(2) $\dfrac{[H^+][NO_2^-]}{[HNO_2]}$ (4) $\dfrac{[H^+][HNO_2]}{[NO_2^-]}$

40 Based on Reference Table L, which of the following 0.1 M aqueous solutions is the best conductor of electricity?
(1) HF (3) HNO_2
(2) H_2S (4) CH_3COOH

41 How many milliliters of 4.00 M NaOH are required to exactly neutralize 50.0 milliliters of a 2.00 M solution of HNO_3?
(1) 25.0 mL   (3) 100. mL
(2) 50.0 mL   (4) 200. mL

42 Which reaction best illustrates amphoterism?
(1) $H_2O + HCl \rightarrow H_3O^+ + Cl^-$
(2) $NH_3 + H_2O \rightarrow NH_4^+ + OH^-$
(3) $H_2O + H_2SO_4 \rightarrow H_3O^+ + HSO_4^-$
(4) $H_2O + H_2O \rightarrow H_3O^+ + OH^-$

43 Given the oxidation-reduction reaction:

$$Hg^{2+} + 2I^- \rightarrow Hg(\ell) + I_2(s)$$

Which equation correctly represents the half-reaction for the reduction that occurs?
(1) $Hg^{2+} \rightarrow Hg(\ell) + 2e^-$
(2) $Hg^{2+} + 2e^- \rightarrow Hg(\ell)$
(3) $2I^- \rightarrow I_2(s) + 2e^-$
(4) $2I^- + 2e^- \rightarrow I_2(s)$

44 In any oxidation-reduction reaction, the total number of electrons gained is
1 less than the total number of electrons lost
2 greater than the total number of electrons lost
3 equal to the total number of electrons lost
4 unrelated to the total number of electrons lost

45 The oxidation number of nitrogen N_2O is
(1) +1   (3) –1
(2) +2   (4) –2

46 When 1 mole of Sn^{4+} ions is reduced to 1 mole of Sn^{2+} ions, 2 moles of electrons are
1 lost by Sn^{4+} 3 gained by Sn^{4+}
2 lost by Sn^{2+} 4 gained by Sn^{2+}

47 The purpose of the salt bridge in an electrochemical cell is to
1 prevent the migration of ions
2 permit the migration of ions
3 provide a direct path for electron flow
4 prevent electron flow

48 Given the reaction:

$$2KCl(\ell) \rightarrow 2K(s) + Cl_2(g)$$

In this reaction, the K^+ ions are
1 reduced by losing electrons
2 reduced by gaining electrons
3 oxidized by losing electrons
4 oxidized by gaining electrons

49 What is the maximum number of covalent bonds that an atom of carbon can form?
(1) 1   (3) 3
(2) 2   (4) 4

50 Which class of organic compounds can be represented as R—OH?
1 acids   3 esters
2 alcohols   4 ethers

51 What is the geometric shape of a methane molecule?
   1 triangular   3 octahedral
   2 rectangular   4 tetrahedral

52 Which molecule contains a total of three carbon atoms?
   (1) 2-methylpropane   (3) propane
   (2) 2-methylbutane   (4) butane

53 Which is the general formula for an alkyne?
   (1) C_nH_{2n-2} (3) C_nH_{2n}
   (2) C_nH_{2n+2} (4) C_nH_{2n-6}

Note that questions 54 through 56 have only three choices.

54 As the temperature of a given sample of gas is increased
   at constant pressure, the volume of the gas will
   1 decrease
   2 increase
   3 remain the same

55 As HCl(aq) is added to a basic solution, the pH of the
   solution will
   1 decrease
   2 increase
   3 remain the same

56 Given the reaction: $A(g) + B(g) \rightarrow C(g)$
   As the concentration of $A(g)$ increases, the frequency of
   collisions of $A(g)$ with $B(g)$
   1 decreases
   2 increases
   3 remains the same

PART 2: *This part consists of twelve groups. Choose seven of these twelve groups. Be sure to answer all questions in each group chosen. Write the answers to these questions on the separate answer sheet provided.* [35]

GROUP 1—Matter and Energy

If you choose this group, be sure to answer questions **57–61**.

57 Which statement describes a characteristic of all compounds?
1 Compounds contain one element, only.
2 Compounds contain two elements, only.
3 Compounds can be decomposed by chemical means.
4 Compounds can be decomposed by physical means.

58 The pressure on 30. milliliters of an ideal gas increases from 760 torr to 1520 torr at constant temperature. The new volume is

(1) $30. \text{ mL} \times \dfrac{760 \text{ torr}}{1520 \text{ torr}}$

(2) $30. \text{ mL} \times \dfrac{1520 \text{ torr}}{760 \text{ torr}}$

(3) $\dfrac{760 \text{ torr}}{30. \text{ mL}} \times 1520 \text{ torr}$

(4) $\dfrac{1520 \text{ torr}}{30. \text{ mL}} \times 760 \text{ torr}$

59 Which phase change results in a release of energy?
(1) $Br_2(\ell) \rightarrow Br_2(s)$ (3) $H_2O(s) \rightarrow H_2O(\ell)$
(2) $I_2(s) \rightarrow I_2(g)$ (4) $NH_3(\ell) \rightarrow NH_3(g)$

60 The *minimum* number of fixed points required to establish the Celsius temperature scale for a thermometer is

(1) 1　　　　　　　　(3) 3
(2) 2　　　　　　　　(4) 4

61 Which substance is a binary compound?

1 oxygen　　　　　　3 glycerol
2 chlorine　　　　　　4 ammonia

GROUP 2—Atomic Structure

If you choose this group, be sure to answer questions 62–66.

62 Which atom in the ground state contains one completely filled *p*-orbital?

(1) N　　　　　　　　(3) He
(2) O　　　　　　　　(4) Be

63 A gamma ray is best described as having

1 no electric charge and no mass
2 a negative charge and no mass
3 a positive charge and a mass number of 2
4 a positive charge and a mass number of 4

64 If one-eighth of the mass of the original sample of a radioisotope remains unchanged after 4,800 years, the isotope could be

(1) ^{3}H　　　　　　　(3) ^{90}Sr
(2) ^{42}K　　　　　　(4) ^{226}Ra

65 An atom of an element has the electron configuration $1s^2 2s^2 2p^2$. What is the total number of valence electrons in this atom?

(1) 6   (3) 5

(2) 2   (4) 4

66 What is the total number of electrons in a Mg^{2+} ion?

(1) 10   (3) 12

(2) 2   (4) 24

GROUP 3—Bonding

*If you choose this group, be sure to answer questions **67–71**.*

67 Given the unbalanced equation:

$$\_\_Ag(s) + \_\_H_2S(g) \rightarrow \_\_Ag_2S(s) + \_\_H_2(g)$$

What is the sum of the coefficients when the equation is completely balanced using the smallest whole-number coefficients?

(1) 5   (3) 10

(2) 8   (4) 4

68 Which ion has the electron configuration of a noble gas?

(1) Cu^{2+} (3) Ca^{2+}

(2) Fe^{2+} (4) Hg^{2+}

69 Which is a molecular substance?

(1) CO_2 (3) KCl

(2) CaO (4) $KClO_3$

70 Given the phase change: $H_2(g) \rightarrow H_2(\ell)$
Which kind of force acts between molecules of H_2 during this phase change?

1 hydrogen bond   3 molecule-ion
2 ionic bond   4 van der Waals

71 Given the electron dot formula: $H \!:\! X \!:\!\!\overset{..}{}$
                                                      H

The attraction for the bonding electrons would be greatest when X represents an atom of

(1) S   (3) Se
(2) O   (4) Te

GROUP 4—Periodic Table

If you choose this group, be sure to answer questions 72–76.

72 A nonmetal could have an electronegativity of

(1) 1.0   (3) 1.6
(2) 2.0   (4) 2.6

73 Which group below contains elements with the greatest variation in chemical properties?

(1) Li, Be, B   (3) B, Al, Ga
(2) Li, Na, K   (4) Be, Mg, Ca

74 A property of most nonmetals in the solid state is that they are

1 good conductors of heat
2 good conductors of electricity
3 brittle
4 malleable

75 A transition element in the ground state could have an electron configuration of

(1) $1s^2 2s^2 2p^6 3s^2 3p^6 4s^2$

(2) $1s^2 2s^2 2p^6 3s^2 3p^6 3d^5 4s^2$

(3) $1s^2 2s^2 2p^6 3s^2 3p^6 3d^{10} 4s^2 4p^5$

(4) $1s^2 2s^2 2p^6 3s^2 3p^6 3d^{10} 4s^2 4p^6$

76 An atom of which of the following elements has the smallest covalent radius?

(1) Li   (3) C

(2) Be   (4) F

GROUP 5—Mathematics of Chemistry

If you choose this group, be sure to answer questions 77–81.

77 Given the reaction: $2PbO \rightarrow 2Pb + O_2$

What is the total volume of O_2, measured at STP, produced when 1.00 mole of PbO decomposes?

(1) 5.60 L   (3) 22.4 L

(2) 11.2 L   (4) 44.8 L

78 What is the mass in grams of 3.0×10^{23} molecules of CO_2?

(1) 22 g   (3) 66 g

(2) 44 g   (4) 88 g

79 Why is salt (NaCl) put on icy roads and sidewalks in the winter?

1 It is ionic and lowers the freezing point of water.

2 It is ionic and raises the freezing point of water.

3 It is covalent and lowers the freezing point of water.

4 It is covalent and raises the freezing point of water.

80 Which gas diffuses most rapidly at STP?
   (1) Ar
   (2) Kr
   (3) N_2
   (4) O_2

81 How many calories of heat are absorbed when 70.00 grams of water is completely vaporized at its boiling point?
   (1) 7,706
   (2) 77.06
   (3) 3,776
   (4) 37,760

GROUP 6—Kinetics and Equilibrium

*If you choose this group, be sure to answer questions **82–86**.*

82 Which reaction system tends to become *less* random as reactants form products?
   (1) $C(s) + O_2(g) \rightarrow CO_2(g)$
   (2) $S(s) + O_2(g) \rightarrow SO_2(g)$
   (3) $I_2(s) + Cl_2(g) \rightarrow 2ICl(g)$
   (4) $2Mg(s) + O_2(g) \rightarrow 2MgO(s)$

83 Given the equilibrium system at 25°C:

   $NH_4Cl(s) \leftrightarrows NH_4^+(aq) + Cl^-(aq)$
   $(\Delta H = +3.5 \text{ kcal/mole})$

   Which change will shift the equilibrium to the right?
   1 decreasing the temperature to 15°C
   2 increasing the temperature to 35°C
   3 dissolving NaCl crystals in the equilibrium mixture
   4 dissolving NH_4NO_3 crystals in the equilibrium mixture

84 What is the correct equilibrium expression (K_{sp}) for the reaction below?

$$Ca_3(PO_4)_2(s) \leftrightharpoons 3Ca^{2+}(aq) + 2PO_4^{3-}(aq)$$

(1) $K_{sp} = [3Ca^{2+}] [2PO_4^{3-}]$
(2) $K_{sp} = [3Ca^{2+}] + [2PO_4^{3-}]$
(3) $K_{sp} = [Ca^{2+}]^3 [PO_4^{3-}]^2$
(4) $K_{sp} = [Ca^{2+}]^3 + [PO_4^{3-}]^2$

85 What is the value of ΔG for any chemical reaction at equilibrium?
1 one
2 zero
3 greater than one
4 less than one but not zero

86 Based on Reference Table M, which of the following salts is the most soluble in water?
(1) $PbCrO_4$ (3) $BaSO_4$
(2) $AgBr$ (4) $ZnCO_3$

GROUP 7—Acids and Bases

If you choose this group, be sure to answer questions 87–91.

87 Which formulas represent a conjugate acid-base pair?
(1) CH_3COOH and CH_3COO^-
(2) H_3O^+ and OH^-
(3) H_2SO_4 and SO_4^{2-}
(4) H_3PO_4 and PO_4^{3-}

88 Which substance, when dissolved in water, is a Brönsted-Lowry acid?
(1) CH_3OH
(2) NaOH
(3) C_2H_5COOH
(4) CH_3COO^-

89 Which of the following compounds is the strongest electrolyte?
(1) NH_3
(2) H_2O
(3) H_3PO_4
(4) H_2SO_4

90 Red litmus will turn blue when placed in an aqueous solution of
(1) HCl
(2) CH_3COOH
(3) CH_3OH
(4) NaOH

91 Given the reaction at equilibrium:

$$HSO_4^- + H_2O \leftrightarrows H_3O^+ + SO_4^{2-}$$

The two Brönsted bases are
(1) H_2O and H_3O^+
(2) H_2O and SO_4^{2-}
(3) H_3O^+ and HSO_4^-
(4) H_3O^+ and SO_4^{2-}

GROUP 8—Redox and Electrochemistry

If you choose this group, be sure to answer questions 92–96.

Base your answers to questions 92 and 93 on Reference Table N and the diagram below.

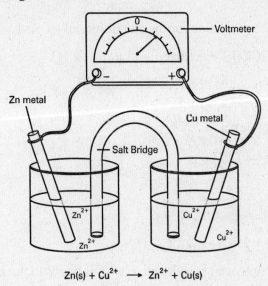

$$Zn(s) + Cu^{2+} \longrightarrow Zn^{2+} + Cu(s)$$

92 When this cell operates, the electrons flow from the
   1 copper half-cell to the zinc half-cell through the wire
   2 copper half-cell to the zinc half-cell through the salt bridge
   3 zinc half-cell to the copper half-cell through the wire
   4 zinc half-cell to the copper half-cell through the salt bridge

93 What is the potential (E^0) of this cell?
   (1) +1.10 V   (3) –1.10 V
   (2) +0.42 V   (4) –0.42 V

94 Based on Reference Table N, which ion will reduce Ag^+ to Ag?
(1) F^- (3) Br^-
(2) I^- (4) Cl^-

95 Given the equation for the electrolysis of a fused salt:

$$2LiCl(\ell) + electricity \rightarrow 2Li(\ell) + Cl_2(g)$$

Which reaction occurs at the cathode?
(1) $2Cl^- \rightarrow Cl_2(g) + 2e^-$
(2) $2Cl^- + 2e^- \rightarrow Cl_2(g)$
(3) $Li^+ + e^- \rightarrow Li(\ell)$
(4) $Li^+ \rightarrow Li(\ell) + e^-$

96 Given the reaction:

$$Zn + 2HCl \rightarrow ZnCl_2 + H_2$$

Which statement best describes what happens to the zinc?
1 The oxidation number changes from +2 to 0, and the zinc is reduced.
2 The oxidation number changes from 0 to +2, and the zinc is reduced.
3 The oxidation number changes from +2 to 0, and the zinc is oxidized.
4 The oxidation number changes from 0 to +2, and the zinc is oxidized.

GROUP 9—Organic Chemistry

If you choose this group, be sure to answer questions 97–101.

97 Given the structural formulas for three alcohols:

$$H-\underset{\underset{H}{|}}{\overset{\overset{H}{|}}{C}}-\underset{\underset{H}{|}}{\overset{\overset{H}{|}}{C}}-OH, \quad H-\underset{\underset{H}{|}}{\overset{\overset{H}{|}}{C}}-\underset{\underset{OH}{|}}{\overset{\overset{H}{|}}{C}}-\underset{\underset{H}{|}}{\overset{\overset{H}{|}}{C}}-H, \quad H-\underset{\underset{H}{|}}{\overset{\overset{H}{|}}{C}}-\underset{\underset{OH}{|}}{\overset{\overset{\overset{\overset{H}{|}}{C}}{|}}{\overset{H}{|}}}-\underset{\underset{H}{|}}{\overset{\overset{H}{|}}{C}}-H$$

All are classified as

1 monohydroxy alcohols   3 tertiary alcohols
2 secondary alcohols   4 primary alcohols

98 Which structural formula correctly represents 2-butene?

(1) $H-\overset{\overset{H}{|}}{C}=\overset{\overset{H}{|}}{C}-\underset{\underset{H}{|}}{\overset{\overset{H}{|}}{C}}-\underset{\underset{H}{|}}{\overset{\overset{H}{|}}{C}}-H$

(2) $H-\underset{\underset{H}{|}}{\overset{\overset{H}{|}}{C}}-\overset{\overset{H}{|}}{C}=\overset{\overset{H}{|}}{C}-\underset{\underset{H}{|}}{\overset{\overset{H}{|}}{C}}-H$

(3) $H-C\equiv C-\underset{\underset{H}{|}}{\overset{\overset{H}{|}}{C}}-\underset{\underset{H}{|}}{\overset{\overset{H}{|}}{C}}-H$

(4) $H-\underset{\underset{H}{|}}{\overset{\overset{H}{|}}{C}}-C\equiv C-\underset{\underset{H}{|}}{\overset{\overset{H}{|}}{C}}-H$

99 Which compounds are isomers?
(1) CH_3Br and CH_2Br_2
(2) CH_3OH and CH_3CH_2OH
(3) CH_3OH and CH_3CHO
(4) CH_3OCH_3 and CH_3CH_2OH

100 Condensation polymerization is best described as
1 a dehydration reaction
2 a cracking reaction
3 a reduction reaction
4 an oxidation reaction

101 Which type of reaction do ethane molecules and ethene
molecules undergo when they react with chlorine?
1 Ethane and ethene both react by addition.
2 Ethane and ethene both react by substitution.
3 Ethane reacts by substitution and ethene reacts by
addition.
4 Ethane reacts by addition and ethene reacts by
substitution.

GROUP 10—Applications of Chemical Principles
*If you choose this group, be sure to answer questions **102–106**.*

102 Which type of chemical reaction generates the electrical
energy produced by a battery?
1 oxidation-reduction   3 neutralization
2 substitution   4 addition

103 Given the reaction at equilibrium:

$$2SO_2(g) + O_2(g) \leftrightarrows 2SO_3(g) + heat$$

Which change will shift the equilibrium to the right?
1 adding a catalyst
2 adding more $O_2(g)$
3 decreasing the pressure
4 increasing the temperature

104 Which commercial products are derived primarily from petroleum?
1 mineral acids and plastics
2 fertilizers and rubber
3 plastics and textiles
4 textiles and fertilizers

105 Petroleum is classified chemically as
1 a substance
2 a compound
3 an element
4 a mixture

106 Which metal forms a self-protective coating against corrosion?
(1) Fe
(2) Cu
(3) Zn
(4) Mg

GROUP 11—Nuclear Chemistry

If you choose this group, be sure to answer questions 107–111.

107 Given the nuclear equation:

$$^9_4Be + X \rightarrow + ^6_3Li + ^4_2He$$

What is the identity of particle X in this equation?

(1) ^1_1H

(3) $^0_{-1}e$

(2) ^2_1H

(4) ^1_0n

108 Which fields are used in accelerators to speed up charged particles?

1 magnetic fields, only

2 electric fields, only

3 magnetic and electric fields

4 magnetic and gravitational fields

109 Which nuclear equation represents artificial transmutation?

(1) $^{238}_{93}U \rightarrow ^{234}_{90}Th + ^4_2He$

(2) $^{27}_{13}Al + ^4_2He \rightarrow ^{30}_{15}P + ^1_0n$

(3) $^{226}_{88}Ra \rightarrow ^4_2He + ^{222}_{86}Rn$

(4) $^{14}_6C \rightarrow ^{14}_7N + ^0_{-1}e$

110 The number of neutrons available in a fission reactor is adjusted by the

1 moderator

3 shielding

2 control rods

4 coolant

111 Which substance is sometimes used to slow down the neutrons in a nuclear reactor?
- (1) U-233
- (2) Pu-236
- (3) sulfur
- (4) heavy water

GROUP 12—Laboratory Activities

If you choose this group, be sure to answer questions 112–116.

112 Solubility data for salt X is shown in the table below.

Temperature (°C)	Solubility $\left(\dfrac{\text{g salt } X}{100\,\text{g H}_2\text{O}} \right)$
10	5
20	10
30	15
40	20
50	30
60	35

Which graph most closely resembles the data shown in the table?

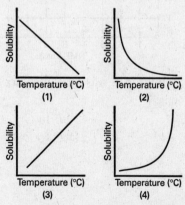

113 A student calculated the percent by mass of water in a hydrate to be 37.2%. If the accepted value is 36.0%, the percent error in the student's calculation is equal to

(1) $\dfrac{1.2}{37.2} \times 100$

(3) $\dfrac{37.2}{36.0} \times 100$

(2) $\dfrac{1.2}{36.0} \times 100$

(4) $\dfrac{36.0}{37.2} \times 100$

114 The graph below represents the cooling curve of a substance starting at a temperature below the boiling point of the substance.

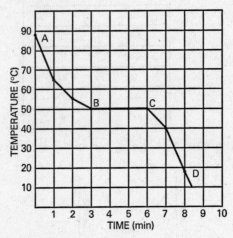

During which interval was the substance completely in the solid phase?

(1) A to B

(3) B to C

(2) A to C

(4) C to D

115 What is the sum of

6.6412 g + 12.85 g + 0.046 g + 3.48 g,

expressed to the correct number of significant figures?
(1) 23 g   (3) 23.017 g
(2) 23.0 g   (4) 23.02 g

116 What occurs as potassium nitrate is dissolved in a beaker of water, indicating that the process is endothermic?
1 The temperature of the solution decreases.
2 The temperature of the solution increases.
3 The solution changes color.
4 The solution gives off a gas.

ANSWER KEY
JUNE 1995

PART 1

1. 3	15. 3	29. 3	43. 2
2. 4	16. 1	30. 4	44. 3
3. 2	17. 2	31. 3	45. 1
4. 2	18. 4	32. 2	46. 3
5. 1	19. 2	33. 1	47. 2
6. 3	20. 3	34. 4	48. 2
7. 4	21. 4	35. 1	49. 4
8. 1	22. 4	36. 3	50. 2
9. 1	23. 1	37. 1	51. 4
10. 3	24. 3	38. 4	52. 3
11. 4	25. 4	39. 2	53. 1
12. 2	26. 1	40. 3	54. 2
13. 1	27. 1	41. 1	55. 1
14. 4	28. 2	42. 4	56. 2

PART 2

57. 3	72. 4	87. 1	102. 1
58. 1	73. 1	88. 3	103. 2
59. 1	74. 3	89. 4	104. 3
60. 2	75. 2	90. 4	105. 4
61. 4	76. 4	91. 2	106. 3
62. 2	77. 2	92. 3	107. 1
63. 1	78. 1	93. 1	108. 3
64. 4	79. 1	94. 2	109. 2
65. 4	80. 3	95. 3	110. 2
66. 1	81. 4	96. 4	111. 4
67. 1	82. 4	97. 1	112. 2
68. 3	83. 2	98. 2	113. 2
69. 1	84. 3	99. 4	114. 4
70. 4	85. 2	100. 1	115. 4
71. 2	86. 3	101. 3	116. 1

ANSWERS AND EXPLANATIONS
JUNE 1995

PART 1

1. **3** According to the reaction

   Cl + Cl → Cl$_2$ + 58 kcal,

   two chlorine atoms react with each other to produce a diatomic chlorine molecule. In addition, 58 kcal (heat) is written as a product (to the right of the arrow). Therefore, heat is released when a bond is formed between two chlorine atoms. A reaction in which heat is a product is called an *exothermic reaction*.

   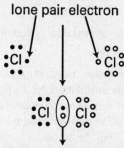

   lone pair electron

   covalent bond is formed between
   two chlorine atoms when their unpaired
   valence electrons are shared.

2. **4** There are three phases of matter: solid, liquid, and gas. Solids have a definite shape and volume. Liquids have no definite shape but possess a definite volume. Liquids take the shape of the container in which they are held. Gases have no definite shape or volume.

3. **2** The prefix *kilo* means 1,000, or 10^3. Therefore, a kilocalorie is equal to 1,000 calories. 1/100 (10^{-2}) of a calorie is equal to a centicalorie and 1/1,000 (10^{-3}) of a calorie is equal to a millicalorie.

4. **2** Chemical decomposition leads to a breakdown of complex compounds into simple units. By definition, a simple unit cannot be decomposed

because it cannot be broken down further. Therefore, an element—the simplest chemical unit—cannot be decomposed. Of the answer choices given, carbon (choice 2) is the only element.

5. 1 When sugar is dissolved in water, the resulting solution is called a *homogeneous mixture* because it is uniform in nature. The mixture has the appearance of being a pure substance.

6. 3 Electrons fill out the various energy levels in the following order: $1s^2 2s^2 2p^6 3s^2 3p^6 4s^2 3d^{10} 4p^6$. According to this electron configuration, electrons must fill out the orbitals exactly in the order they appear. An electron is termed to be in an *excited state* if an electron jumps to the next orbital before the previous orbital is completely filled. In the electron configuration $1s^2 2s^2 2p^5 3s^2$, electrons jumped to the $3s^2$ orbital before the $2p^5$ orbital was completely filled ($2p$ can hold a maximum of six electrons).

7. 4 According to the periodic table, the ground state of lithium has the following electron configuration: $1s^2 2s^1$.

8. 1 In an atom, protons have positive charge and electrons have negative charge. Neutrons are neutral in nature. Therefore, if a particle contains six protons and six electrons, it must also contain six positive and six negative charges. Because there are an equal number of positive and negative charges, the overall charge of the particle must be zero. Therefore, this particle is a neutral atom. Using the periodic table, we can determine that the element with six protons is carbon. As a result, the particle in question must be a neutral carbon atom (choice 1).

9. 1 Because the electrode is negative in charge, it only attracts positive charges to itself. Of the choices given, only alpha particles possess positive charges. Both beta particles and electrons have negative charges; a neutron has no charge.

10. 3 The mass of an electron is $\frac{1}{1,836}$ of a proton. As a result, only protons and neutrons account for the atomic mass. The electron's mass is so minute that it can be ignored when calculating the atomic mass. If you do not know exactly how much smaller the mass of an electron is, compared to a proton, you can still answer this question. You need to remember that the mass of an electron is so small compared to the

mass of a proton that it does not contribute to the mass of an atom. Choices 1 and 2 give the mass of a proton, and choice 4 is only slightly less than the mass of a proton. The only answer that gives a mass significantly less than the mass of a proton is choice 3.

11. 4 The mass number is the sum of the protons and neutrons in the nucleus of an atom.

12. 2 When you are choosing between ionic and molecular substances, you should remember that only ionic substances can conduct electricity. To conduct electricity, particles must be charged and able to move. Based on these two requirements of conductivity, choices 3 and 4 can be ruled out because $C_6H_{12}H_6$ is an organic (molecular) compound. Choice 1 is incorrect because NaCl is in solid phase, and the ions are immobile in the ionic lattice. To conduct electricity, NaCl must be either molten (liquid phase) or dissolved in solution.

13. 1 An empirical formula is the smallest whole number ratio of elements in a specific compound. Of the answer choices given, K_2O (choice 1) cannot be further reduced to a smaller whole number ratio. Therefore, it must represent an empirical formula. H_2O_2 can be reduced to HO, C_2H_6 can be reduced to CH_3, and C_6H_6 can be reduced to CH.

14. 4 To have molecule-ion attractions, there must be a homogenous mixture (solution) containing molecules and ions. Of the four choices given, only choice 4 represents a solution; it is an aqueous solution of an ionic compound that contains both molecules (H_2O) and ions (K^+ and Cl^-). Remember that ionic compounds dissociate in aqueous solution.

15. 3 Choice 3 ($CuCl_2$) is correctly paired with its name, copper (II) chloride. The names of all binary compounds end with the suffix -ide. In addition, Cl has a charge of –1 in $CuCl_2$. Because there are two Cl molecules, the total negative charge in the compound is –2. Therefore, copper must have a charge of +2. The correct name of $MgCl_2$ is magnesium chloride, the correct name of K_2O is potassium oxide, and the correct name of FeO is iron (II) oxide. (Oxygen carries a –2 charge in ionic compounds; the iron carries a charge of +2.)

16. 1 The molecule $C_2H_4(OH)_2$ has six hydrogen atoms. There are four atoms of hydrogen bonded to C_2 and two atoms of hydrogen in the

(OH)$_2$ portion of the compound. Therefore, in one molecule of C$_2$H$_4$(OH)$_2$, there are six H atoms, and in 1 mole of C$_2$H$_4$(OH)$_2$ there are 6 moles of H atoms.

17. 2 A coordinate bond is formed when a single element contributes both electrons to a bond formed between two nonmetallic elements. In the case of bond A, the two electrons shared by N and H are donated by the nitrogen. The polyatomic ion NH$_4^+$ has a total of eight valence electrons. As indicated in the diagram, five of the electrons (denoted by an x) originate from the nitrogen (you can determine the number of valence electrons on N by looking at the periodic table). The three electrons denoted with a small O originate from three of the hydrogen atoms. In bond A, both of the electrons being shared between H and N are x's and originate from the nitrogen. Therefore, choice 2 correctly describes the source of the two electrons in bond A.

18. 4 A polar molecule is the result of an overall unequal distribution of electrons within the molecule. If a molecule possesses either a lone pair (unshared pair) or lone pairs of electrons on the central atom, then the molecule automatically has an unequal distribution of charge and is considered polar. NH$_3$ has a lone pair on the nitrogen so it is polar (see diagram). Choice 1 contains no polar bonds and is perfectly symmetric. Choices 2 and 3 are perfectly symmetric molecules with no lone pairs of electrons on the central atom, and so they are nonpolar.

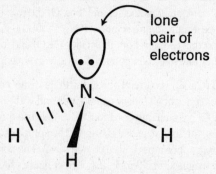

19. 2 Metalloids are found on either side of the bold line that separates metals from nonmetals on the periodic table. Of the choices given, only choice 2 (Sb) lies next to the bold line that separates metals from nonmetals.

GROUPS

13	14	15	16	17
B	C	N	O	F
Al	Si	P	S	Cl
Ga	Ge	As	Se	Br
In	Sn	Sb	Te	I
Tl	Pb	Bi	Po	At

20. **3** Good electrical conductivity and luster are characteristics of metals. As we move from top to bottom within a group on the periodic table, the metallic characteristics increase. Bi (choice 3) lies at the bottom of group 15. Therefore, Bi has the best luster and electric conductivity of the four choices given. It is also the only element that does not lie to the right or is adjacent to the bold line which separates metals and nonmetals.

21. **4** The period number of an element is the number of its outermost principal energy level. Therefore, the electron configuration of an element in period 3 has 3 as its highest energy level (or principal quantum number). Of the four electron configurations given, only choice 4 ($1s^2 2s^2 2p^6 3s^1$) indicates that an electron is present in energy level 3.

22. **4** Refer to Reference Table P for a list of atomic radii of all elements. According to Table P, strontium has the largest covalent radius. In general, atomic radii or covalent radii increase going down a group and decrease going left to right across a period.

23. **1** Metals have both low ionization energies and low electronegativity values. As a result, it is easier for metals to lose electrons from their valence shells. Therefore, when metals react with nonmetals, metals lose electrons to form positively charged ions (cations).

24. 3 The alkaline earth metals are found in group 2 of the periodic table. Group 2 elements are metallic in nature. In addition, group 2 elements have two electrons in their valence shells. When these elements lose the two electrons to form an ion (and gain a noble gas electron configuration), they acquire a +2 charge.

25. 4 The compound with the fewest ionic characteristics has the smallest difference in electronegativity values between its atoms. Refer to Reference Table K to solve this problem. By using the values stated in Table K, we find that the difference in electronegativity between C and Cl in CCl_4 is calculated as $0.6(3.2 - 2.6)$. The electronegative difference of KCl is 2.4, the electronegative difference of $CaCl_2$ is 2.2, and the electronegative difference of $AlCl_3$ is 1.7.

26. 1 Because Na (sodium) has a low ionization energy and a low electronegativity, sodium can easily react in nature to form stable ionic compounds. In nature, Na often reacts with elements such as chlorine and oxygen. As a result, it is difficult to find Na in its uncombined state.

   You can also answer this question by using a logical argument. You have two decisions to make. First, is Na reactive or unreactive? If Na is unreactive then it would not readily form compounds and Na would be abundant. We are told that Na is not abundant; therefore, Na is *reactive*. Second, does Na form stable or unstable compounds? If Na forms unstable compounds, then these compounds would break down to form Na. We already know that Na is not abundant; therefore, Na forms *stable* compounds.

27. 1 The gram molecular mass of calcium nitrate, $Ca(NO_3)_2$, is

Ca	40 g (atomic mass) $\times 1$ = 40 g
N	14 g (atomic mass) $\times 2$ = 28 g
O	16 g (atomic mass) $\times 6$ = 96 g
	164 g

   Be careful of those parentheses!

28. 2 Molarity = moles of solute/volume of solution in liters
   Moles of solute = mass of solute in grams/formula mass of solute

Mass of solute in grams is 116 g.

Formula mass of KF is (39 g + 19 g) = 58 g

Moles of solute = 116 g/58 g = 2.00 mol

$$116 \text{ g} \times \frac{1 \text{ mol}}{58 \text{ g}}$$

Molarity = 2 moles of solute/1.00 liter of solution = 2.00 M

29. **3** One mole of any substance contains 6×10^{23} particles (Avogadro's number). Therefore, a mole of calcium also contains the same number of particles—6×10^{23} atoms.

30. **4** % water = $\dfrac{\text{mass of water}}{\text{formula mass}} \times 100\%$

$$= \frac{18 \text{ g}}{\text{mol}} \times \frac{10 \text{ mol H}_2\text{O}}{1 \text{ mol Na}_2\text{CO}_3 \cdot 10\text{H}_2\text{O}} = \frac{180 \text{ g H}_2\text{O}}{1 \text{ mol Na}_2\text{CO}_3 \cdot 10\text{H}_2\text{O}}$$

Mass of water = 18 g per molecule × 10 mol = 180 g

Formula mass = 286 g/mol

Percent of water by mass = 180 g/286 g × 100% = 62.9%

31. **3** Thirty-two liters of O_2 contains the same number of molecules as 32.0 liters of H_2. According to Avogadro's hypothesis, at the same temperature and volume (in liters), all gases contain the same number of molecules. Because the volumes of both O_2 and H_2 are the same, they must also contain the same number of molecules.

32. **2** Reference Table D shows the solubility curves of various compounds. According to Table D, as temperature rises from 10°C to 70°C, the solubility of NH_3 decreases more rapidly than the other three choices.

33. **1** Reactants react with each other to form products. When reactants are consumed, a reaction can no longer form products. As a result, the rate of chemical reaction decreases. Reaction rate is also dependent on the probability of atoms colliding with each other. The higher the collision rate, the greater the reaction rate. When temperature is decreased, the average kinetic energy of atoms is lowered. As a result, they move slowly. Slow-moving molecules collide with less frequency. The result of a lowered frequency of collisions is that the chemical reaction rate is lowered.

34. 4 At equilibrium, the rate of the forward and the reverse reactions are equal to each other. Choice 4 correctly states the definition of a dynamic equilibrium. Be careful not to quickly choose choice 2. Recall that it is only the *ratio* of products to reactants that remains constant at equilibrium. That ratio can be any value and is *not* necessarily 1:1; in other words, there are not equal amounts of reactants and products at equilibrium.

35. 1 A spontaneous reaction has a negative Gibb's free energy (ΔG°_f). Consult Reference Table G, which lists the Gibb's free energy of various compounds. All of the reactions given are formation reactions. Of the four answer choices given, only the reaction in choice 1 has a negative energy of formation (ΔG°_f). All other choices have positive ΔG°_f values.

36. 3 Refer to Reference Table D to answer this question. According to Table D, the solubility of KCl increases as temperature of the solvent increases. Stirring and increasing surface area of the solute only changes the rate at which KCl dissolves but does not *increase the amount* of KCl that is dissolved. Pressure has no affect on KCl because it is a solid. Pressure only affects compounds that are in the gaseous phase.

37. 1 An Arrhenius acid is an acid that has the ability to release H^+ ions in aqueous solution. Choice 3 (OH^-) indicates a base. Choices 2 and 4 have nothing to do with acids and bases.

38. 4 Neutralization reactions produce water (H_2O) when H^+ and OH^- combine with each other. Choice 4 correctly shows the formation of water through a neutralization reaction. All other choices show acid-base reactions leading to formation of a new acid or a base.

39. 2 The ionization constant (K_a) is calculated by dividing the equilibrium concentration of the products of the reaction by the equilibrium concentration of the reactants.

$$K_a = [products]/[reactants]$$

$$= \frac{[H^+]\,[NO_2^-]}{[HNO_2]}$$

40. 3 Reference Table L lists the ionization or acid dissociation constants (K_a) of various acids. As the ionization constant increases, the ratio of

ions produced to acid dissolved in solution increases. Therefore, the greater the ionization constant, the greater the concentration of ions. As the concentration of ions increases, the conductivity of a solution increases. Therefore, acids found at the top of Table L are better conductors of electricity. According to Table L, HNO_2 (choice 3) has the largest ionization constant. Therefore, it is the best conductor of electricity in the group.

41   1   When a solution reaches neutrality, the number of moles of acid is equal to the number of moles of base (moles of acid = moles of base). Based on this fundamental, we can use the following formula to calculate the volume of NaOH:

$$Volume_{(base)} \times molarity_{(base)} = volume_{(acid)} \times molarity_{(acid)}$$

$$Volume_{(base)} \times 4.00 \text{ M} = 50.0 \text{ ml} \times 2.00 \text{ M}$$

$$Volume_{(base)} = 25.0 \text{ ml}$$

42. 4   An amphoteric species acts as both an acid and a base. You can easily identify an amphoteric substance. All amphoteric substances have a negative charge and at least one hydrogen atom as part of their structures. The only exception to this rule is the water molecule. The water molecule can act as a base and an acid.

43. 2   When a species is reduced, it gains electrons. In other words, the oxidation number of the species decreases when it undergoes reduction. Of the four answer choices given, only choice 2 illustrates reduction. In the reaction, Hg^{2+} gains two electrons to become Hg^0.

44. 3   Reduction-oxidation, or redox, reactions involve simultaneous loss and gain of electrons. In a redox reaction, one of the reactants loses electrons whereas the second reactant picks up those electrons, and the total number of electrons gained is equal to the total number of electrons lost.

45. 1   The overall charge of N_2O is zero. However, we know that the oxidation number, or charge, of oxygen is –2. This means that all the nitrogen in N_2O must have a +2 charge to balance the –2 charge of the oxygen. Because there are two nitrogen atoms in the compound, each nitrogen must have a +1 charge ($+1 \times 2 = +2$).

46. 3   When a mole of Sn^{4+} is reduced to Sn^{2+}, 2 moles of electrons are gained by Sn^{4+}. Recall that during reduction, oxidation number is lowered. In this case, Sn changed from +4 to +2. This means that two negative charges (two electrons) must have been added to the Sn^{4+} atom: $Sn^{4+} + 2e^- \rightarrow Sn^{2+}$. The total charge on the left side of the equation equals the total charge on the right side of the equation. Choices 1 and 2 are not correct because an atom gains and does not lose electrons during reduction. Choice 4 is incorrect because Sn^{2+} is not being reduced.

47. 2   An electrochemical cell or battery requires a continuous flow of charged particles (electrons or ions). A salt bridge is necessary to form a complete circuit. A salt bridge allows ions to migrate from one half-cell to the other.

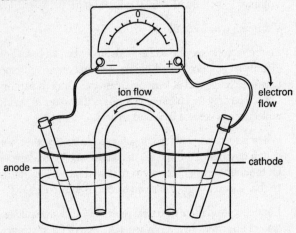

48. 2   KCl is an ionic compound (K is a metal, and Cl is a nonmetal). In an ionic compound, the metal has a positive charge. In this case, K (potassium) has a charge of +1 (K^+). In the given reaction, KCl decomposes to form elemental K and Cl_2. All elements have a charge of zero. Therefore, the oxidation number of K is reduced from +1 to 0. In other words, K must have undergone reduction by gaining electrons.

49. 4   Carbons form a maximum of four covalent bonds. Carbon has a total of four valence electrons ($2s^2 2p^2$), and each one of those electrons can

pair up with an electron from another atom to form a covalent bond, which completes carbon's octet.

50. **2** The functional group —OH represents alcohols. Organic acids are represented by the functional group —COOH, esters are represented by the functional group —COO—, and ethers are represented by the functional group —COC—.

$$R - \overset{\overset{\displaystyle O}{\displaystyle \|}}{C} - O - H \qquad \text{Carboxylic acid}$$

$$R_1 - \overset{\overset{\displaystyle O}{\displaystyle \|}}{C} - O - R_2 \qquad \text{Ester}$$

$$R_1 - O - R_2 \qquad \text{Ether}$$

51. **4** Methane (CH_4) is a symmetrical molecule in which carbon forms a single bond with each hydrogen atom. Because there are no lone (unpaired) *pairs* of electrons, the molecule assumes a three-dimensional tetrahedral shape in which carbon is located in the center and each hydrogen is located at the corners of the tetrahedron.

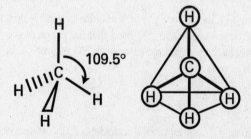

all angles are 109.5°

52. **3** The prefix *prop*- indicates a chain of three carbons. Choice 3 (propane) has a molecular formula of C_3H_8.

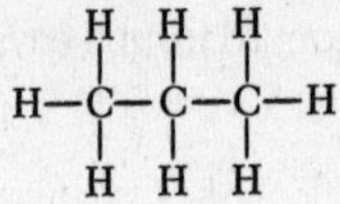

Propane

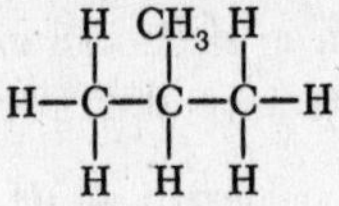

2-Methylpropane

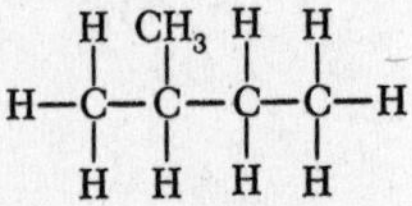

2-Methylbutane

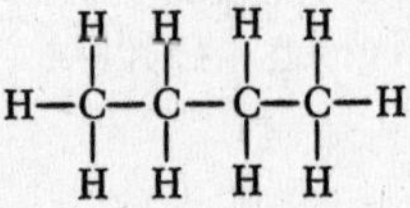

Butane

53. 1 An alkyne is a hydrocarbon molecule that has a triple bond in it. The general formula for an alkyne is C_nH_{2n-2}. C_nH_{2n+2} is a general formula for alkanes (single bonds), and C_nH_2 is the general formula for alkenes (double bonds).

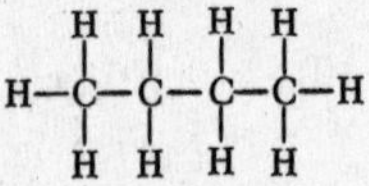

alkane—only single bonds are present—C_nH_{2n+2}

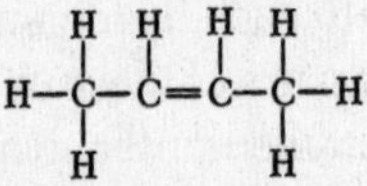

alkene—a double bond is present—C_nH_{2n}

54. 2 According to Charles' law, the temperature of a gas is directly proportional to the volume of the gas, given that the pressure is constant. Therefore, as the temperature increases, the volume of the gas increases as well.

$$\frac{V_1}{T_1} = \frac{V_2}{T_2}$$

55. 1 HCl is an acid. In solution, HCl dissociates to its component ions H^+ and Cl^-. When the concentration of H^+ increases in solution, the pH decreases because the solution becomes more acidic. Therefore when HCl is added to a basic solution, the pH of the solution decreases (becomes more acidic).

56. **2** As the concentration of A(g) increases, the total number of particles of A also increases. Therefore, the probability of A colliding with B increases. The greater the number of particles present, the greater the possibility that they will collide with each other.

PART 2

Group 1—Matter and Energy

57. **3** Compounds are composed of two or more elements bonded through chemical bonds. Choices 1 and 2 are incorrect because compounds can be of two or more elements. Choice 4 is also incorrect because mixtures and not compounds can be separated by physical means.

58. **1** The relationship between pressure and volume in gases at constant temperature is stated by Boyle's law: $P_1 \cdot V_1 = P_2 \cdot V_2$

$V_2 = V_1 \cdot P_1/P_2$

$= 30 \text{ ml} \cdot 760 \text{ torr}/1520 \text{ torr}$

59. **1** An exothermic reaction releases energy on completion. During phase change, energy is released when a substance changes from a gas to a liquid and from a liquid to a solid. Therefore, the correct answer is choice 1 because it depicts a liquid changing to a solid.

$Br_2(\ell) \rightarrow Br_2(s)$

It is often easier to recall that it takes energy to melt a solid sample, and therefore the reverse process (freezing) must release energy.

60. **2** The minimum number of fixed points necessary to establish the Celsius temperature scale for a thermometer is two. Because the Celsius scale is divided into 100 equal parts, the scale requires a minimum and a maximum point. The minimum scale on the Celsius scale is 0°C (freezing point of water), and the maximum point is 100°C (boiling point of water). With these two fixed points, the size of the Celsius scale is established.

61. **4** The word *binary* means two. Therefore, a binary compound is composed of two elements. Of the four choices given, only ammonia (NH_3) contains two elements—nitrogen and hydrogen. Choices 1

(oxygen) and 2 (chlorine) are elements. They are not compounds. Choice 3 (glycerol) is composed of three elements—carbon, hydrogen, and oxygen. Be careful not to confuse binary with diatomic. Binary indicates that a compound is composed of two different elements in any whole number ratio. A diatomic molecule is composed of a total of two atoms, which can be of the same element (e.g., O_2 and Cl_2 are diatomic but not binary).

Group 2—Atomic Structure

62. **2** To have one completely filled p orbital, there must be at least four electrons present in the p subshell. In its ground state, oxygen has an electron configuration of $1s^2 2s^2 2p^4$. In an oxygen atom, the first p orbital is filled (see diagram). Helium and beryllium lack electrons in their p orbitals. Nitrogen has three electrons in the p orbital, so it lacks the minimum of four electrons needed to completely fill one p orbital.

$$\frac{\uparrow\downarrow}{p_x} \quad \frac{\uparrow}{p_y} \quad \frac{\uparrow}{p_z}$$

Location of four electrons in p orbitals of oxygen

63. **1** Gamma radiation belongs to the electromagnetic spectrum. In other words, it behaves like light. Light has no electric charge or mass. Therefore, gamma radiation also lacks any electric charge or mass.

64. **4** The half-life of a radioactive substance is the time required for the substance to decay to one-half its present value. Therefore, if ⅛ of a substance remains unchanged after 4,800 years, then it must have been through 3 half-lives ($\frac{1}{2}^3 = \frac{1}{8}$). The half-life of the substance in question must be 1,600 years (4,800/3 = 1,600 years). According to Reference Table H, the isotope ^{226}Ra has a half-life of 1,600 years.

$$100\% \xrightarrow[\text{half-life}]{\text{one}} 50\% \xrightarrow[\text{half-life}]{\text{one}} 25\% \xrightarrow[\text{half-life}]{\text{one}} 12.5\%$$

65. 4 The total number of valence electrons in an atom is equal to the number of electrons present in the electron configuration's highest energy level. An atom with the electron configuration of $1s^22s^22p^2$ has a total of four valence electrons. Notice that there are exactly four electrons in the second energy level ($2s^22p^2$).

66. 1 The atomic number of magnesium (Mg) is 12. That means a magnesium atom has 12 protons and 12 electrons. When a magnesium atom ionizes to Mg^{2+}, two electrons are lost by the atom. Therefore, Mg^{2+} has a total of 10 electrons.

Group 3—Bonding

67. 1 The reaction, when balanced, has the following mole ratio:

$$2Ag(s) + 1H_2S(g) \rightarrow 1Ag_2S(g) + 1H_2$$

Therefore, the sum of the coefficients in the balanced equation is $2 + 1 + 1 + 1 = 5$.

68. 3 Each of the choices is an ion with a +2 charge. This means that in forming the ions, each neutral atom lost two electrons through reduction. The electron configuration of Ca^{2+} is the same as the configuration of the noble gas argon: $Ca^{2+} = 1s^22s^22p^63s^23p^6$. The electron configuration of the ions in all other choices does not match the electron configuration of any noble gases:

Choice 1: Cu^{2+} [Ar] $4s^23d^7$

Choice 2: Fe^{2+} [Ar] $4s^13d^5$

Choice 4: Hg^{2+} [Xe] $5d^{10}$

69. 1 Molecular substances have covalent bonds in them. Covalent bonds are created when two nonmetals bind to each other. Of the four answer choices given, only choice 1 (CO_2) is a covalent compound. All other choices are ionic in nature because they are composed of a metal and a nonmetal (or, in the case of choice 4, a metal and two nonmetals—$KClO_3^-$ is a polyatomic ion).

70. 4 Only weak forces exist between H_2 molecules. H_2 is a nonpolar molecule, and, therefore, attractive forces such as hydrogen-bond, ionic-bond, or molecule-ion forces cannot exist between H_2 molecules. Of

the four answer choices given, only van der Waals is a weak force (van der Waals forces are often called *dispersion forces*).

71. 2 The attraction for the bonding electrons would be greatest when X represents an atom with a high electronegativity. Of the four elements given, oxygen (choice 2) has the highest electronegativity. Use Reference Table K to determine the electronegativity of elements.

Group 4—Periodic Table

72. 4 Nonmetals have high electronegativity. Of the answer choices given, the most likely electronegativity related to nonmetals is the one that is the highest in a group on the periodic table. The highest electronegativity of the four answer choices is 2.6. From Reference Table K, we can also determine that sulfur (S), a nonmetal, has an electronegativity of 2.6.

73. 1 The greatest variation in chemical properties is observed as you move from left to right within a period. On the other hand, the least variation is observed as you move from top to bottom within a group. The correct answer is choice 1 because the elements Li, Be, and B all belong to the same period. Therefore, the greatest chemical variation is seen among them.

74. 3 Nonmetals are characterized by the following physical properties: Nonmetals are brittle and poor conductors of electricity and heat.

75. 2 The transitional elements are located between groups 2 and 11 in the periodic table. The only electron configuration that fits a transitional element is $1s^2 2s^2 2p^6 3s^2 3p^6 3d^5 4s^2$. This is the electron configuration of the transitional element manganese (Mn). The d subshell is not filled, which indicates that the element is a transition metal.

76. 4 The covalent or atomic radii decrease from left to right across a period and from bottom to top in a group. According to Reference Table P, fluorine (F) has the smallest covalent radius (0.64 angstroms).

Group 5—Mathematics of Chemistry

77. 2 According to the balanced reaction $2PbO \rightarrow 2Pb + O_2$, for every 2 moles of PbO that decomposes, 1 mole of O_2 is produced. Therefore,

if only 1 mole of PbO reacts, then half a mole of O_2 is produced. One mole of all gases, including O_2, occupies 22.4 liters. Therefore, half a mole of O_2 occupies 11.2 liters.

78. 1 One mole of any substance has 6.0×10^{23} molecules. Therefore, 3.0×10^{23} molecules are present in half a mole of a substance. One mole of CO_2 contains 44 g of mass. Half a mole of CO_2 contains 2 g of mass.

$$
\begin{array}{lll}
C & 1 \times 12 = 12 \\
+ \ O & 2 \times 16 = 32 \\
\hline
& \quad\quad 44 \text{ g/mol}
\end{array}
$$

$$3.0 \times 10^{23} \text{ molecules} \times \frac{1 \text{ mol}}{6.0 \times 10^{23} \text{ molecules}} \times \frac{44 \text{ g CO}_2}{1 \text{ mol}}$$

$$= 22 \text{ g CO}_2$$

79. 1 When solutes are added to a solvent (liquid), the freezing point is depressed and the boiling point is elevated. One factor that determines how much the freezing point is lowered is the number of particles (solute) present in the water. This is called the *colligative property* of a solvent. An ionic compound like NaCl, when put on icy roads and sidewalks in the winter, lowers the freezing point of water.

80. 3 According to Graham's law, at STP the gas with the lowest molar mass diffuses most rapidly. In other words, lighter gases diffuse faster than heavier gases. Of the four gases given, nitrogen (N_2) is the lightest. Therefore, it diffuses more rapidly than the other three gases.

81. 4 The heat of vaporization is the amount of heat needed to vaporize a given mass of liquid at its boiling point. The heat of vaporization of water is 539.4 calories per gram.

q (number of calories of heat) = mass of water × heat of vaporization
$$= 70.00 \text{ g} \times 539.4 \text{ cal/g}$$
$$= 37,760 \text{ cal}$$

Group 6—Kinetics and Equilibrium

82. 4 There are three phases of matter: solid, liquid, and gas. The solid state is the most ordered state, and the gaseous state is the most ran-

dom (disordered state). Therefore, any reaction that produces a solid as a product is the least random (the most ordered). According to answer choice 4, a solid (MgO) is produced at the conclusion of the reaction. Of the four answer choices given, the least random product is produced in choice 4.

83. **2** Enthalpy (ΔH) tells us the amount of heat released or absorbed in a reaction. If ΔH is positive, heat is consumed as a reactant in the reaction. The Le Châtelier principle states that when a system in dynamic equilibrium is subjected to a disturbance that upsets the equilibrium, the system undergoes a change that counterbalances the disturbance. Disturbances that cause the equilibrium to shift to the right in this problem are either adding a reactant or removing a product. Only choice 2 fits these criteria because increasing the temperature can be thought of as adding a reactant. Choice 1 removes a reactant, choice 3 adds a product [$Cl^-(aq)$], and choice 4 adds a product [$NH^{4+}(aq)$].

84. **3** The solubility constant (K_{sp}) is calculated as the product of the equilibrium concentrations of all products of a reaction raised to the power of their coefficients in the balanced equation. Therefore, the K_{sp} of the following reaction is

$$Ca_3(PO_4)_2 \rightleftarrows 3Ca^{2+} + 2PO_4^{3-}$$

$$K_{sp} = [Ca^{2+}]^3 [PO_4^{3-}]^2$$

85. **2** At equilibrium, ΔG is equal to zero. Remember that at equilibrium the rates of forward and reverse reactions are equal. The reaction is non-spontaneous in either direction. When ΔG is negative, a reaction is spontaneous, and when ΔG is positive, a reaction is nonspontaneous.

86. **3** Reference Table M lists the solubility products (K_{sp}) of various compounds. The larger the K_{sp}, the more soluble a compound is in solution. Conversely, the smaller the K_{sp}, the less soluble a compound is in solution. Of the four compounds given, $BaSO_4$ (choice 3) has the highest K_{sp}. Therefore, it is the most soluble in water.

Group 7—Acids and Bases

87. **1** A conjugate acid-base pair differs in structure by a *single* proton (H^+). In choice 1, CH_2COOH and CH_3COO^- differ from each other by a

single proton. In all other choices, the acids and bases differ by more than a proton. Therefore, they cannot be acid-base conjugate pairs.

88. **3** All organic acids have the functional group —COOH. Only one answer choice bears the functional group —COOH. Choice 3 correctly depicts an acid. A Brönsted-Lowry acid must be able to donate a proton. Choice 1 is an alcohol, choice 2 is a base, and choice 4 is the conjugate base of acetic acid.

89. **4** A strong electrolyte produces ions in solution. The stronger an electrolyte, the more ions it produces when it is dissolved in solution. By comparing K_a and K_b values in Reference Tables L and M, we can determine that H_2SO_4 is the strongest electrolyte.

NH_3: K_b = 1.8×10^{-5}
H_2O: K_w = 1.0×10^{-4}
H_3PO_4: K_a = 3×10^{-2}
H_2SO_4: K_a = large; completely dissociates in aqueous solution

90. **4** Red litmus turns blue in a basic solution, such as NaOH. Choices 1 and 2 (HCl and CH_3COOH) are acids, and choice 3 (CH_3OH) is an alcohol.

91. **2** Brönsted acids donate a proton (H^+), and Brönsted bases accept a proton (H^+). To determine the two Brönsted bases, determine the acid-base conjugate pairs in the given reaction:

$HSO_4^- + H_2O \rightarrow H_3O^+ + SO_4^{2-}$
acid   base   acid   base

HSO_4^- and SO_4^{2-} are a conjugate acid-base pair, as are H_3O^+ and H_2O. Therefore, the two Brönsted bases are SO_4^{2-} and H_2O.

Group 8—Redox and Electrochemistry

92. **3** In an electrochemical cell, electrons flow from the electrode where oxidation takes place to the electrode where reduction occurs. In the given battery, oxidation takes place at the Zn electrode and reduction occurs at the Cu electrode. Therefore, electrons flow from the zinc half-cell to the copper half-cell through the wire.

93. **1** First, determine the species that undergoes reduction and the species that undergoes oxidation. In the given reaction, Zn undergoes oxidation, and Cu^{2+} undergoes reduction. Next, determine the electrode potentials of Zn and Cu from Reference Table N.

$$Zn = -0.76 \text{ V}$$
$$Cu = +0.34 \text{ V}$$

Finally, find the cell voltage (E^0) by subtracting the electrode potential of the species being oxidized from that of the species that is being reduced.

$$E^0 = E^0{}_{reduced} - E^0{}_{oxidized}$$
$$= (+0.34 \text{ V}) - (-0.76 \text{ V}) = +1.10 \text{ V}$$

94. 2 The ion that reduces Ag^+ to Ag must have a lower reduction potential than Ag^+ because the ion itself must undergo oxidation. Reference Table N lists in decreasing order the reduction potential of various elements and ions. Therefore, an ion that can reduce Ag^+ must be below Ag^+ in Table N

95. 3 The cathode is the electrode where reduction occurs. In an electrolytic cell, the negative electrode is the cathode. It attracts the positive ions to it and supplies the electrons for their reduction. Therefore, the reduction of Li^+ ions occurs at the cathode.

96. 4 In the reaction $Zn + 2HCl \rightarrow ZnCl_2 + H_2$, the zinc's oxidation number changes from zero to +2. Because the zinc lost two electrons, it underwent oxidation. The half-reaction looks like this:

$$Zn^0 \rightarrow Zn^{2+} + 2e^-$$

The charge on the left equals the total charge on the right side of the equation: $0 = +2 + -2 = 0$.

Group 9—Organic Chemistry

97. 1 Alcohols are characterized by the functional group —OH. Of the three alcohols given, all three structures contain a single —OH functional group. Therefore, they must be classified as *monohydroxy* alcohols (the prefix *mono* means one).

98. 2 A butene is an alkene. All alkenes contain at least one double bond. Therefore, choices 3 and 4 cannot be the answer because they contain triple bonds. In 2-butene, the double bond must be present between the second and third carbons. The number 2 indicates the carbon where the double bond originates. Therefore, choice 2 is the correct answer.

99. 4 Isomers have the same molecular formula but different structural arrangements. Only choice 4 has two molecules whose molecular formula is the same (same number of carbon, oxygen, and hydrogen) but whose structures are different. All the other choices have pairs of molecules with different molecular formulas.

100. 1 By definition, a condensation reaction occurs when two molecules react with each other and in the process a molecule of water is eliminated between them (H_2O is a product molecule). Loss of a water molecule is also called *dehydration*. Therefore, condensation polymerization is best described as a dehydration reaction.

101. 3 Ethane is an alkane, whereas ethene is an alkene. Alkanes are saturated molecules where all the carbons are in a single bond with other atoms. On the other hand, alkenes are unsaturated molecules because double bonds are present. Therefore, when ethane reacts with chlorine, a substitution reaction occurs (see diagram). However, when ethene reacts with chlorine, the chlorine adds across the double bond. This type of reaction is called an *addition reaction*.

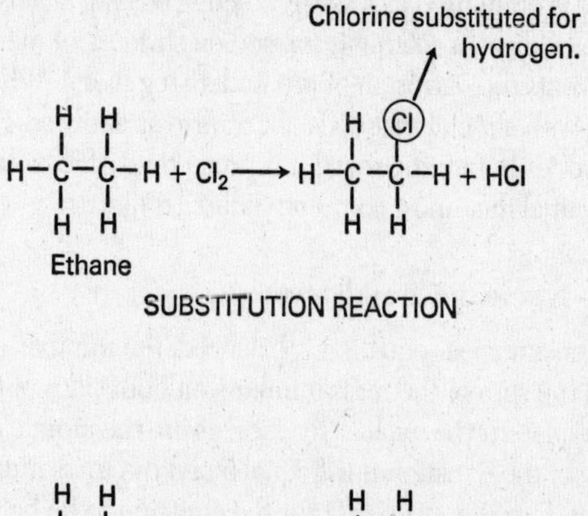

Chlorine substituted for hydrogen.

Ethane

SUBSTITUTION REACTION

Ethene

The two chlorine atoms add to the ethene molecule.

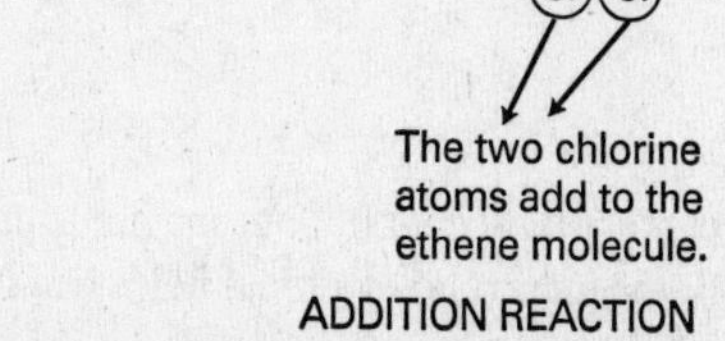

ADDITION REACTION

Group 10—Applications of Chemical Principles

102. 1   All electrochemical batteries produce electricity by generating moving electrons. Electrons are generated through reduction-oxidation (redox) reactions.

103. 2   Because O_2 is a reactant in the given reaction, increasing the amount of O_2 shifts the equilibrium to the right. Catalysts only speed up reactions but do not affect equilibrium. Decreasing the pressure shifts the equilibrium to the left. Because this reaction is an exothermic reaction, addition of more heat favors the reverse reaction. Thus the equilibrium shifts to the left in response to increased temperature.

104. 3   Plastics and textiles are polymers, or chains of molecules, made of hydrocarbon molecules. Petroleum products are also polymers of similar hydrocarbons.

105. 4   Because petroleum is composed of polymers of hydrocarbons and other compounds, petroleum is classified as a mixture.

106. 3   Zinc is often used as a self-protecting coating against corrosion. When zinc oxidizes, a coating is formed on surfaces of metals. This coating prevents the corrosion of the underlying metal. When zinc is in contact with a metal that does not oxidize as easily as zinc, the zinc always corrodes before the metal it is protecting. Zinc has a lower reduction potential than most common industrial metals.

Group 11—Nuclear Chemistry

107. 1   When a nuclear equation is balanced, the number of atomic numbers and the sum of the mass numbers on both sides of the equation must be equal. In the nuclear reaction given, the atomic number on each side of the equation must equal 5 and the mass number must be 10. Therefore, X must be ^1_1H for the equation to be balanced.

$$^9_4Be + \,^1_1H \rightarrow \,^6_3Li + \,^4_2He$$

$$9 + 1 = 6 + 4$$

$$4 + 1 = 3 + 2$$

108. 3   Charged particles are deflected in both magnetic and electrical fields. Therefore, both magnetic and electrical fields can be used to accelerate charged particles.

109. 2 Artificial transmutation involves the experimental or artificial bombardment of a nucleus with another nuclear particle. As a result, an element is transformed into another element. The reaction in choice 2 is an example of artificial transmutation. Choices 1 and 3 are examples of alpha decay, and choice 2 is an example of beta decay.

110. 2 The control rods absorb excess neutrons in fission reactors. By regulating the number of neutrons they absorb, control rods can be used to adjust the number of available neutrons.

111. 4 Heavy water is used as a moderator in reactors. U-233 and Pu-236 are used as fuel. Sulfur is not an important substance in the function of nuclear reactors.

Group 12—Laboratory Activities

112. 2 According to the trend observed in the data table, as temperature increases, solubility increases as well. Closer inspection of the data also reveals that the proportion by which both temperature and solubility increase is exactly the same. Each time the temperature is raised 10°C, solubility increases by 5 g per 100 g of H_2O. Graphs 1 and 3 both indicate that the solubility decreases with increasing temperature. Graph 4 indicates that the solubility increases exponentially with increasing temperature (it looks like one-half of a parabola). Only graph 3 shows the solubility increasing linearly with temperature.

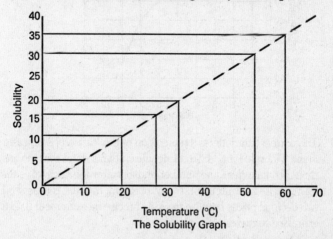

Temperature (°C)
The Solubility Graph

113. **2** Percent error can be calculated with the following formula:

$$\% \text{ error} = \frac{\text{experimental value} - \text{accepted value}}{\text{accepted value}} \times 100\%$$

$$= \frac{37.2 - 36.0 \times 100\%}{36.0} = 1.2/36.0 \times 100\%$$

Be careful not to confuse % error with % yield.

114. **4** Because the substance started at a temperature below the boiling point, the initial phase of the substance is liquid. To be completely in the solid phase, the substance must freeze first. When a substance freezes, there is no change in the temperature. Therefore, freezing must have occurred between points B and C (no change in tempera- ture). The substance must be completely in solid phase between the points C and D.

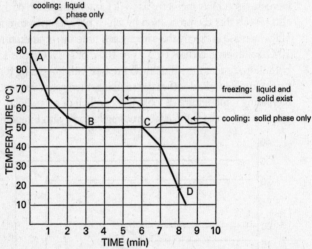

115. **4** The correct sum reflects the precision of the least precise measure- ment. The rule for addition of significant digits states that the answer should contain the same number of digits after the decimal as the least precise measurement. In this problem, the sum is expressed to two decimal places because the least precise measurement 3.48 is expressed to two decimal places.

116. 1 Endothermic processes absorb energy from their surroundings. Therefore, when potassium nitrate is dissolved in a beaker of water, the temperature of the solution decreases. The dissolving process extracts heat from water to carry out the endothermic process, thus lowering the temperature of the solution.

EXAMINATION
JUNE 1996

PART 1: *Answer all 56 questions in this part.* [65]

DIRECTIONS (1–56): *For each statement or question, select the word or expression that, of those given, best completes the statement or answers the question. Record your answer on the separate answer sheet provided.*

---

1 What is the vapor pressure of a liquid at its normal boiling temperature?
   (1) 1 atm   (3) 273 atm
   (2) 2 atm   (4) 760 atm

2 A sealed container has 1 mole of helium and 2 moles of nitrogen at 30°C. When the total pressure of the mixture is 600 torr, what is the partial pressure of the nitrogen?
   (1) 100 torr   (3) 400 torr
   (2) 200 torr   (4) 600 torr

3 Solid X is placed in contact with solid Y. Heat will flow spontaneously from X to Y when
   (1) X is 20°C and Y is 20°C
   (2) X is 10°C and Y is 5°C
   (3) X is –25°C and Y is –10°C
   (4) X is 25°C and Y is 30°C

4 Which graph represents the relationship between volume and Kelvin temperature for an ideal gas at constant pressure?

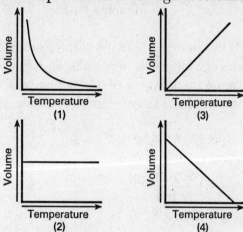

5 An example of a binary compound is
(1) potassium chloride
(2) ammonium chloride
(3) potassium chlorate
(4) ammonium chlorate

6 Which kind of radiation will travel through an electric field on a pathway that remains unaffected by the field?
(1) a proton   (3) an electron
(2) a gamma ray   (4) an alpha particle

7 The major portion of an atom's mass consists of
(1) electrons and protons
(2) electrons and neutrons
(3) neutrons and positrons
(4) neutrons and protons

8 Which atom contains exactly 15 protons?
   (1) phosphorus-32   (3) oxygen-15
   (2) sulfur-32   (4) nitrogen-15

9 Element X has two isotopes. If 72.0% of the element has
   an isotopic mass of 84.9 atomic mass units, and 28.0% of
   the element has an isotopic mass of 87.0 atomic mass
   units, the average atomic mass of element X is numeri-
   cally equal to
   (1) $(72.0 + 84.9) \times (28.0 + 87.0)$
   (2) $(72.0 - 84.9) \times (28.0 + 87.0)$
   (3) $\dfrac{(72.0 - 84.9)}{100} + \dfrac{(28.0 + 87.0)}{100}$
   (4) $(72.0 \times 84.9) + (28.0 \times 87.0)$

10 Given the equation: $^{14}_{6}C \rightarrow ^{14}_{7}N + X$
    Which particle is represented by the letter X?
    (1) an alpha particle   (3) a neutron
    (2) a beta particle   (4) a proton

11 The atom of which element in the ground state has 2
    unpaired electrons in the $2p$ sublevel?
    (1) fluorine   (3) beryllium
    (2) nitrogen   (4) carbon

12 Which atoms contain the same number of neutrons?
    (1) $^{1}_{1}H$ and $^{3}_{2}He$ (3) $^{3}_{1}H$ and $^{3}_{2}He$

    (2) $^{2}_{1}H$ and $^{4}_{2}He$ (4) $^{3}_{1}H$ and $^{4}_{2}He$

13 Which hydrocarbon formula is also an empirical formula?
    (1) CH_4 (3) C_3H_6
    (2) C_2H_4 (4) C_4H_8

14 The potential energy possessed by a molecule is dependent upon
  (1) its composition, only
  (2) its structure, only
  (3) both its composition and its structure
  (4) neither its composition nor its structure

15 Which is a correctly balanced equation for a reaction between hydrogen gas and oxygen gas?
  (1) $H_2(g) + O_2(g) \rightarrow H_2O(\ell) + heat$
  (2) $H_2(g) + O_2(g) \rightarrow 2H_2O(\ell) + heat$
  (3) $2H_2(g) + 2O_2(g) \rightarrow H_2O(\ell) + heat$
  (4) $2H_2(g) + O_2(g) \rightarrow 2H_2O(\ell) + heat$

16 The atom of which element has an ionic radius smaller than its atomic radius?
  (1) N   (3) Br
  (2) S   (4) Rb

17 Which molecule contains a polar covalent bond?

  (1) ×× ••
       ×I× :I: (3)   H ×• N •× H
       ×× ••   H

  (2) H ×• H   (4)   :N ו×N ×

18 Which is the correct formula for nitrogen (I) oxide?
  (1) NO   (3) NO_2
  (2) N_2O (4) N_2O_3

19 Which element in Group 15 has the strongest metallic character?

(1) Bi   (3) P

(2) As   (4) N

20 Which halogens are gases at STP?
   (1) chlorine and fluorine
   (2) chlorine and bromine
   (3) iodine and fluorine
   (4) iodine and bromine

21 What is the total number of atoms represented in the formula $CuSO_4 \cdot 5H_2O$?

(1) 8   (3) 21

(2) 13   (4) 27

22 When combining with nonmetallic atoms, metallic atoms generally will
   (1) lose electrons and form negative ions
   (2) lose electrons and form positive ions
   (3) gain electrons and form positive ions
   (4) gain electrons and form positive ions

23 Which set of elements contains a metalloid?

(1) K, Mn, As, Ar   (3) Ba, Ag, Sn, Xe

(2) Li, Mg, Ca, Kr   (4) Fr, F, O, Rn

24 Atoms of elements in a group on the Periodic Table have similar chemical properties. This similarity is most closely related to the atoms'
(1) number of principal energy levels
(2) number of valence electrons
(3) atomic numbers
(4) atomic masses

25 Which element in Period 2 of the Periodic Table is the most reactive nonmetal?
(1) carbon   (3) oxygen
(2) nitrogen   (4) fluorine

26 What is the gram formula mass of $(NH_4)_3PO_4$?
(1) 113 g   (3) 149 g
(2) 121 g   (4) 404 g

27 Given the reaction:

$$CH_4 + 2O_2 \rightarrow CO_2 + 2H_2O$$

What amount of oxygen is needed to completely react with 1 mole of CH_4?
(1) 2 moles   (3) 2 grams
(2) 2 atoms   (4) 2 molecules

28 Based on Reference Table E, which of the following saturated solutions would be the *least* concentrated?
(1) sodium sulfate
(2) potassium sulfate
(3) copper (II) sulfate
(4) barium sulfate

29 What is the total number of moles of H_2SO_4 needed to prepare 5.0 liters of a 2.0 M solution of H_2SO_4?
(1) 2.5   (3) 10.
(2) 5.0   (4) 20.

30 Given the reaction:

$$Ca(s) + 2H_2O(\ell) \rightarrow Ca(OH)_2(aq) + H_2(g)$$

When 40.1 grams of Ca(s) reacts completely with the water, what is the total volume, at STP, of $H_2(g)$ produced?
(1) 1.00 L   (3) 22.4 L
(2) 2.00 L   (4) 44.8 L

31 Which is the correct equilibrium expression for the reaction below?

$$4NH_3(g) + 7O_2(g) \rightleftarrows 4NO_2(g) + 6H_2O(g)$$

(1) $K = \dfrac{[NO_2][H_2O]}{[NH_3][O_2]}$ (3) $K = \dfrac{[NH_3][O_2]}{[NO_2][H_2O]}$

(2) $K = \dfrac{[NO_2]^4[H_2O]^6}{[NH_3]^4[O_2]^7}$ (4) $K = \dfrac{[NH_3]^4[O_2]^7}{[NO_2]^4[H_2O]^6}$

32 The potential energy diagram below shows the reaction $X + Y \rightleftarrows Z$.

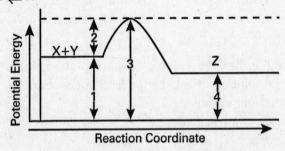

When a catalyst is added to the reaction, it will change the value of

(1) 1 and 2   (3) 2 and 3
(2) 1 and 3   (4) 3 and 4

33 Which conditions will increase the rate of a chemical reaction?
(1) decreased temperature and decreased concentration of reactants
(2) decreased temperature and increased concentration of reactants
(3) increased temperature and decreased concentration of reactants
(4) increased temperature and increased concentration of reactants

34 A solution exhibiting equilibrium between the dissolved and undissolved solute must be
(1) saturated   (3) dilute
(2) unsaturated   (4) concentrated

35 Which 0.1 M solution has a pH greater than 7?
(1) $C_6H_{12}O_6$ (3) KCl
(2) CH_3COOH (4) KOH

36 What color is phenolphthalein in a basic solution?
(1) blue   (3) yellow
(2) pink   (4) colorless

37 According to Reference Table *L*, which of the following is the strongest Brönsted-Lowry acid?

(1) HS^- (3) HNO_2
(2) H_2S (4) HNO_3

38 When HCl(aq) is exactly neutralized by NaOH(aq), the hydrogen ion concentration in the resulting mixture is

(1) always less than the concentration of the hydroxide ions
(2) always greater than the concentration of the hydroxide ions
(3) always equal to the concentration of the hydroxide ions
(4) sometimes greater and sometimes less than the concentration of the hydroxide ions

39 If 20. milliliters of 4.0 M NaOH is exactly neutralized by 20. milliliters of HCl, the molarity of the HCl is

(1) 1.0 M (3) 5.0 M
(2) 2.0 M (4) 4.0 M

40 The value of the ionization constant of water, K_w, will change when there is a change in

(1) temperature
(2) pressure
(3) hydrogen ion concentration
(4) hydroxide ion concentration

41 Based on Reference Table *L*, which species is amphoteric?

(1) NH_2^- (3) I^-
(2) NH_3 (4) HI

42 A redox reaction is a reaction in which
   (1) only reduction occurs
   (2) only oxidation occurs
   (3) reduction and oxidation occur at the same time
   (4) reduction occurs first and then oxidation occurs

43 Given the reaction:

$$\_Mg + \_Cr^{3+} \rightarrow \_Mg^{2+} + \_Cr$$

When the equation is correctly balanced using smallest whole numbers, the sum of the coefficients will be
   (1) 10   (2) 7   (3) 5   (4) 4

44 Oxygen has an oxidation number of –2 in
   (1) O_2 (3) Na_2O_2
   (2) NO_2 (4) OF_2

45 Given the statements:
   A   The salt bridge prevents electrical contact between solutions of half-cells.
   B   The salt bridge prevents direct mixing of one half-cell solution with the other.
   C   The salt bridge allows electrons to migrate from one half-cell to the other.
   D   The salt bridge allows ions to migrate from one half-cell to the other.

   Which two statements explain the purpose of a salt bridge used as part of a chemical cell?
   (1) A and C   (3) C and D
   (2) A and D   (4) B and D

46 When a substance is oxidized, it
(1) loses protons
(2) gains protons
(3) acts as an oxidizing agent
(4) acts as a reducing agent

47 In the reaction $Cu + 2Ag^+ \rightarrow Cu^{2+} + 2Ag$, the oxidizing agent is
(1) Cu   (3) Ag^+
(2) Cu^{2+} (4) Ag

48 A compound that is classified as organic must contain the element
(1) carbon   (3) oxygen
(2) nitrogen   (4) hydrogen

49 Which substance is a product of a fermentation reaction?
(1) glucose   (3) ethanol
(2) zymase   (4) water

50 Which of the following hydrocarbons has the *lowest* normal boiling point?
(1) ethane   (3) butane
(2) propane   (4) pentane

51 What type of reaction is

$$CH_3CH_3 + Cl_2 \rightarrow CH_3CH_2Cl + HCl?$$

(1) an addition reaction
(2) a substitution reaction
(3) a saponification reaction
(4) an esterification reaction

52 Which compound is a saturated hydrocarbon?
   (1) ethane   (3) ethyne
   (2) ethene   (4) ethanol

Note that questions 53 through 56 have only three choices.

53 As atoms of elements in Group 16 are considered in
   order from top to bottom, the electronegativity of each
   successive element
   (1) decreases
   (2) increases
   (3) remains the same

54 As the pressure of a gas at 760 torr is changed to 380
   torr at constant temperature, the volume of the gas
   (1) decreases
   (2) increases
   (3) remains the same

55 Given the change of phase: $CO_2(g) \rightarrow CO_2(s)$
   As $CO_2(g)$ changes to $CO_2(s)$, the entropy of the system
   (1) decreases
   (2) increases
   (3) remains the same

56 In heterogeneous reactions, as the surface area of the
   reactants increases, the rate of the reaction
   (1) decreases
   (2) increases
   (3) remains the same

PART 2: *This part consists of twelve groups. Choose seven of these twelve groups. Be sure to answer all the questions in each group chosen. Write the answers to these questions on the separate answer sheet provided.* [35]

GROUP 1—Matter and Energy

*If you choose this group, be sure to answer questions **57–61**.*

57 What is the total number of calories of heat energy absorbed by 15 grams of water when it is heated from 30.°C to 40.°C?
(1) 10.
(3) 25
(2) 15
(4) 150

58 The graph below represents the uniform cooling of a sample of a substance, starting with the substance as a gas above its boiling point.

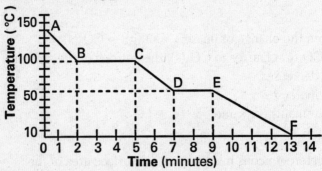

Which segment of the curve represents a time when both the liquid and the solid phases are present?
(1) *EF*
(3) *CD*
(2) *BC*
(4) *DE*

59 Which change of phase is exothermic?
   (1) $NaCl(s) \rightarrow NaCl(\ell)$
   (2) $CO_2(s) \rightarrow CO_2(g)$
   (3) $H_2O(\ell) \rightarrow H_2O(s)$
   (4) $H_2O(\ell) \rightarrow H_2O(g)$

60 According to the kinetic theory of gases, which assumption is correct?
   (1) Gas particles strongly attract each other.
   (2) Gas particles travel in curved paths.
   (3) The volume of gas particles prevents random motion.
   (4) Energy may be transferred between colliding particles.

61 A compound differs from a mixture in that a compound always has a
   (1) homogeneous composition
   (2) maximum of two components
   (3) minimum of three components
   (4) heterogeneous composition

GROUP 2—Atomic Structure

If you choose this group, be sure to answer questions 62–66.

62 An ion with 5 protons, 6 neutrons, and a charge of 3+ has an atomic number of
   (1) 5   (3) 8
   (2) 6   (4) 11

63 Electron X can change to a higher energy level or a lower energy level. Which statement is true of electron X?
(1) Electron X emits energy when it changes to a higher energy level.
(2) Electron X absorbs energy when it changes to a higher energy level.
(3) Electron X absorbs energy when it changes to a lower energy level.
(4) Electron X neither emits nor absorbs energy when it changes energy level.

64 What is the highest principal quantum number assigned to an electron in an atom of zinc in the ground state?
(1) 1   (3) 5
(2) 2   (4) 4

65 The first ionization energy of an element is 176 kilocalories per mole of atoms. An atom of this element in the ground state has a total of how many valence electrons?
(1) 1   (3) 3
(2) 2   (4) 4

66 What is the total number of occupied s orbitals in an atom of nickel in the ground state?
(1) 1   (3) 3
(2) 2   (4) 4

GROUP 3—Bonding

If you choose this group, be sure to answer questions 67–71.

67 What is the chemical formula for nickel (II) hypochlorite?
   (1) $NiCl_2$ (3) $NiClO_2$
   (2) $Ni(ClO)_2$ (4) $Ni(ClO)_3$

68 Based on Reference Table G, which of the following compounds is most stable?
   (1) $CO(g)$ (3) $NO(g)$
   (2) $CO_2(g)$ (4) $NO_2(g)$

69 The attractions that allow molecules of krypton to exist in the solid phase are due to
   (1) ionic bonds
   (2) covalent bonds
   (3) molecule-ion forces
   (4) van der Waals forces

70 Oxygen, nitrogen, and fluorine bond with hydrogen to form molecules. These molecules are attracted to each other by
   (1) ionic bonds
   (2) hydrogen bonds
   (3) electrovalent bonds
   (4) coordinate covalent bonds

71 An atom of which of the following elements has the greatest ability to attract electrons?
   (1) silicon   (3) nitrogen
   (2) sulfur   (4) bromine

GROUP 4—Periodic Table

If you choose this group, be sure to answer questions 72–76.

72 Which electron configuration represents the atom with the largest covalent radius?
   (1) $1s^1$ (3) $1s^2 2s^2$
   (2) $1s^2 2s^1$ (4) $1s^2 2s^2 2p^1$

73 A solution of $Cu(NO_3)_2$ is colored because of the presence of the ion
   (1) Cu^{2+} (3) O^{2-}
   (2) N^{5+} (4) NO_3^{1-}

74 Which element is more reactive than strontium?
   (1) potassium   (3) iron
   (3) calcium   (4) copper

75 At STP, which substance is the best conductor of electricity?
   (1) nitrogen   (3) sulfur
   (2) neon   (4) silver

76 The oxide of metal X has the formula XO. Which group in the Periodic Table contains metal X?
   (1) Group 1   (3) Group 13
   (2) Group 2   (4) Group 17

GROUP 5—Mathematics of Chemistry

*If you choose this group, be sure to answer questions **77–81**.*

77 Given the same conditions of temperature and pressure, which noble gas will diffuse most rapidly?

(1) He   (2) Ne   (3) Ar   (4) Kr

78 What is the total number of molecules of hydrogen in 0.25 mole of hydrogen?

(1) 6.0×10^{23} (3) 3.0×10^{23}

(2) 4.5×10^{23} (4) 1.5×10^{23}

79 The volume of a 1.00-mole sample of an ideal gas will decrease when the

(1) pressure decreases and the temperature decreases

(2) pressure decreases and the temperature increases

(3) pressure increases and the temperature decreases

(4) pressure increases and the temperature increases

80 A 0.100-molal aqueous solution of which compound has the *lowest* freezing point?

(1) $C_6H_{12}O_6$ (3) $C_{12}H_{22}O_{11}$

(2) CH_3OH (4) NaOH

81 What is the empirical formula of a compound that contains 85% Ag and 15% F by mass?

(1) AgF   (2) Ag_2F (3) AgF_2 (4) Ag_2F_2

GROUP 6—Kinetics and Equilibrium

If you choose this group, be sure to answer questions 82–86.

82 Based on Reference Table M, which compound is less soluble in water than $PbCO_3$ at 298 K and 1 atmosphere?
(1) AgI   (2) AgCl   (3) $CaSO_4$ (4) $BaSO_4$

83 Given the equilibrium reaction at constant pressure:

$$2HBr(g) + 17.4 \text{ kcal} \rightleftarrows H_2(g) + Br_2(g)$$

When the temperature is increased, the equilibrium will shift to the
(1) right, and the concentration of HBr(g) will decrease
(2) right, and the concentration of HBr(g) will increase
(3) left, and the concentration of HBr(g) will decrease
(4) left, and the concentration of HBr(g) will increase

84 A system is said to be in a state of dynamic equilibrium when the
(1) concentration of products is greater than the concentration of reactants
(2) concentration of products is the same as the concentration of reactants
(3) rate at which products are formed is greater than the rate at which reactants are formed
(4) rate at which products are formed is the same as the rate at which reactants are formed

85 Which reaction will occur spontaneously? [Refer to Reference Table *G*.]

(1) $\frac{1}{2}N_2(g) + \frac{1}{2}O_2(g) \rightarrow NO(g)$

(2) $\frac{1}{2}N_2(g) + O_2(g) \rightarrow NO(g)$

(3) $2C(s) + 3H_2(g) \rightarrow C_2H_6(g)$

(4) $2C(s) + 2H_2(g) \rightarrow C_2H_4(g)$

86 Which potential energy diagram represents the reaction $A + B \rightarrow C$ + energy?

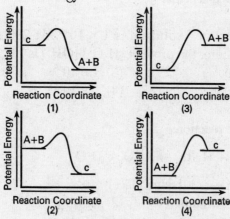

GROUP 7—Acids and Bases

If you choose this group, be sure to answer questions 87–91.

87 Potassium chloride, KCl, is a salt derived from the neutralization of a

(1) weak acid and a weak base

(2) weak acid and a strong base

(3) strong acid and a weak base

(4) strong acid and a strong base

88 Given the reaction:

$$HSO_4^- + H_2O \rightleftarrows H_3O^+ + SO_4^{2-}$$

Which is a Brönsted-Lowry conjugate acid-base pair?
(1) HSO_4^- and H_3O^+ (3) H_2O and SO_4^{2-}
(2) HSO_4^- and SO_4^{2-} (4) H_2O and HSO_4^-

89 An aqueous solution that has a hydrogen ion concentration of 1.0×10^{-8} mole per liter has a pH of
(1) 6, which is basic (3) 8, which is basic
(2) 6, which is acidic (4) 8, which is acidic

90 The $[OH^-]$ of a solution is 1×10^{-6}. At 298 K and 1 atmosphere, the product $[H_3O^+]$ $[OH^-]$ is
(1) 1×10^{-2} (3) 1×10^{-8}
(2) 1×10^{-6} (4) 1×10^{-14}

91 Given the reaction:

$$KOH + HNO_3 \rightarrow KNO_3 + H_2O$$

Which process is taking place?
(1) neutralization (3) substitution
(2) esterification (4) addition

GROUP 8—Redox and Electricity
If you choose this group, be sure to answer questions 92–96.

92 Given the unbalanced equation:

$$\_MnO_2 + \_HCl \rightarrow \_MnCl_2 + \_H_2O + \_Cl_2$$

When the equation is correctly balanced using smallest whole-number coefficients, the coefficient of HCl is
(1) 1   (2) 2   (3) 3   (4) 4

93 Based on Reference Table N, which half-cell has a lower electrode potential than the standard hydrogen half-cell?
(1) $Au^{3+} + 3e^- \rightarrow Au(s)$
(2) $Hg^{2+} + 2e^- \rightarrow Hg(\ell)$
(3) $Cu^+ + e^- \rightarrow Cu(s)$
(4) $Pb^{2+} + 2e^- \rightarrow Pb(s)$

94 According to Reference Table N, which reaction will take place spontaneously?
(1) $Ni^{2+} + Pb(s) \rightarrow Ni(s) + Pb^{2+}$
(2) $Au^{3+} + Al(s) \rightarrow Au(s) + Al^{3+}$
(3) $Sr^{2+} + Sn(s) \rightarrow Sr(s) + Sn^{2+}$
(4) $Fe^{2+} + Cu(s) \rightarrow Fe(s) + Cu^{2+}$

95 Given the reaction:

$$Mg(s) + Zn^{2+}(aq) \rightarrow Mg^{2+}(aq) + Zn(s)$$

What is the cell voltage (E^0) for the overall reaction?
(1) +1.61 V
(2) –1.61 V
(3) +3.13 V
(4) –3.13 V

96 The diagram below represents a chemical cell at 298 K.

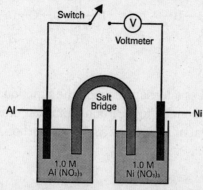

$$2Al(s) + 3Ni^{2+}(aq) \longrightarrow 2Al^{3+} + 3Ni(s)$$

When the switch is closed, electrons flow from

(1) Al(s) to Ni(s)
(2) Ni(s) to Al(s)
(3) Al^{3+}(aq) to Ni^{2+}(aq)
(4) Ni^{2+}(aq) to Al^{3+}(aq)

GROUP 9—Organic Chemistry

If you choose this group, be sure to answer questions 97–101.

97 The compound C_4H_{10} belongs to the series of hydrocarbons with the general formula

(1) C_nH_{2n} (3) C_nH_{2n-2}
(2) C_nH_{2n+2} (4) C_nH_{2n-6}

98 Which is an isomer of

$$\begin{array}{cc} H & H \\ | & | \\ H-C-C-OH \\ | & | \\ H & H \end{array} ?$$

(1)
$$\begin{array}{cc} H & H \\ | & | \\ H-C-O-C-H \\ | & | \\ H & H \end{array}$$

(3)
$$\begin{array}{cc} H & O \\ | & \| \\ H-C-C-H \\ | \\ H \end{array}$$

(2)
$$\begin{array}{cc} H & H \\ | & | \\ HO-C-C-H \\ | & | \\ H & H \end{array}$$

(4)
$$\begin{array}{ccc} H & H & H \\ | & | & | \\ H-C-C-O-C-H \\ | & | & | \\ H & H & H \end{array}$$

99 To be classified as a tertiary alcohol, the functional —OH group is bonded to a carbon atom that must be bonded to a total of how many additional carbon atoms?

(1) 1   (2) 2   (3) 3   (4) 4

100 Which substance is made up of monomers joined together in long chains?
(1) ketone
(3) ester
(2) protein
(4) acid

101 What is the total number of carbon atoms in a molecule of glycerol?
(1) 1   (2) 2   (3) 3   (4) 4

GROUP 10   Applications of Chemical Principles
If you choose this group, be sure to answer questions 102–106.

102 Which type of reaction is occurring when a metal undergoes corrosion?
(1) oxidation-reduction
(2) neutralization
(3) polymerization
(4) saponification

103 Which process is used to separate the components of a petroleum mixture?
(1) addition polymerization
(2) condensation polymerization
(3) fractional distillation
(4) fractional crystallization

104 Which substance functions as the electrolyte in an automobile battery?
(1) PbO_2
(3) H_2SO_4
(2) $PbSO_4$
(4) H_2O

105 A battery consists of which type of cells?
- (1) electrolytic
- (2) electrochemical
- (3) electroplating
- (4) electromagnetic

106 Which element can be found in nature in the free (uncombined) state?
- (1) Ca
- (2) Ba
- (3) Au
- (4) Al

GROUP 11—Nuclear Chemistry

If you choose this group, be sure to answer questions **107–111**.

107 Which radioactive isotope is used in geological dating?
- (1) uranium-238
- (2) iodine-131
- (3) cobalt-60
- (4) technetium-99

108 Which equation represents a fusion reaction?
- (1) $^3_1H + ^1_1H \rightarrow ^4_2He$
- (2) $^{40}_{18}Ar + ^1_1H \rightarrow ^{40}_{19}K + ^0_1n$
- (3) $^{234}_{91}Pa \rightarrow ^{234}_{92}U + ^0_1e$
- (4) $^{226}_{88}Ra \rightarrow ^{226}_{86}Rn + ^4_2He$

109 Which substance is used as a coolant in a nuclear reactor?
- (1) neutrons
- (2) plutonium
- (3) hydrogen
- (4) heavy water

110 Which substance has chemical properties similar to those of radioactive ^{235}U?
- (1) ^{235}Pa
- (2) ^{233}Pa
- (3) ^{233}U
- (4) ^{206}Pb

111 Control rods in nuclear reactors are commonly made
     of boron and cadmium because these two elements
     have the ability to
     (1) absorb neutrons
     (2) emit neutrons
     (3) decrease the speed of neutrons
     (4) increase the speed of neutrons

GROUP 12—Laboratory Activities

If you choose this group, be sure to answer questions **112–116.**

   Base your answers to questions 112 and 113 on the table
below, which represents the production of 50 milliliters of CO_2 in
the reaction of HCl with $NaHCO_3$. Five trials were performed
under different conditions as shown. (The same mass of
$NaHCO_3$ was used in each trial.)

Trial	Particle Size of $NaHCO_3$	Concentration of HCl	Temperature (°C) of HCl
A	small	1 M	20
B	large	1 M	20
C	large	1 M	40
D	small	2 M	40
E	large	2 M	40

112 Which two trials could be used to measure the effect
     of surface area?
     (1) trials A and B   (3) trials A and D
     (2) trials A and C   (4) trials B and D

113 Which trial would produce the fastest reaction?
    (1) trial *A*   (3) trial *C*
    (2) trial *B*   (4) trial *D*

114 A student determined the heat of fusion of water to be
    88 calories per gram. If the accepted value is 80. calo-
    ries per gram, what is the student's percent error?
    (1) 8.0%   (2) 10.%   (3) 11%   (4) 90.%

115 Given: (52.6 cm)(1.214 cm)
    What is the product expressed to the correct number
    of significant figures?
    (1) 64 cm^2 (3) 63.86 cm^2
    (2) 63.9 cm^2 (4) 63.8564 cm^2

116 The diagram below represents a metal bar and two
    centimeter rulers, *A* and *B*. Portions of the rulers have
    been enlarged to show detail.

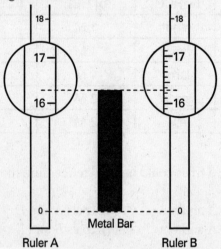

What is the greatest degree of precision to which the metal bar can be measured by ruler A and by ruler B?

(1) to the nearest tenth by both rulers

(2) to the nearest hundredth by both rulers

(3) to the nearest tenth by ruler A and to the nearest hundredth by ruler B

(4) to the nearest hundredth by ruler A and to the nearest tenth by ruler B

ANSWER KEY
JUNE 1996

PART 1

1. 1	15. 4	29. 3	43. 1
2. 3	16. 4	30. 3	44. 2
3. 2	17. 3	31. 2	45. 4
4. 3	18. 2	32. 3	46. 4
5. 1	19. 1	33. 4	47. 3
6. 2	20. 1	34. 1	48. 1
7. 4	21. 3	35. 4	49. 3
8. 1	22. 2	36. 2	50. 1
9. 3	23. 1	37. 4	51. 2
10. 2	24. 2	38. 3	52. 1
11. 4	25. 4	39. 4	53. 1
12. 4	26. 3	40. 1	54. 2
13. 1	27. 1	41. 2	55. 1
14. 3	28. 4	42. 3	56. 2

PART 2

57. 4	72. 2	87. 4	102. 1
58. 4	73. 1	88. 2	103. 3
59. 3	74. 1	89. 3	104. 3
60. 4	75. 4	90. 4	105. 2
61. 1	76. 2	91. 1	106. 3
62. 1	77. 1	92. 4	107. 1
63. 2	78. 4	93. 4	108. 1
64. 4	79. 3	94. 2	109. 4
65. 2	80. 4	95. 1	110. 3
66. 4	81. 1	96. 1	111. 1
67. 2	82. 1	97. 2	112. 1
68. 2	83. 1	98. 1	113. 4
69. 4	84. 4	99. 3	114. 2
70. 2	85. 3	100. 2	115. 2
71. 3	86. 2	101. 3	116. 3

ANSWERS AND EXPLANATIONS JUNE 1996

PART 1

1. **1** Boiling point is defined as the temperature at which the vapor pressure of a liquid is equal to the atmospheric pressure. A boiling point measured at standard atmospheric pressure is called the *normal boiling point*. Standard atmospheric pressure is 1 atm, or 760 mm Hg (torr). Choice 4 is incorrect because the unit of the answer choice is wrong.

2. **3** This is a problem of partial pressures. To solve the problem, solve for the mole fraction of each one of the gases present in the mixture.

   Mole fraction of N_2 = moles of N_2/total moles of all gases in the mixture

   $$= \frac{2 \text{ mol } N_2}{2 \text{ mol } N_2 + 1 \text{ mol He}} = \frac{2 \text{ mol}}{3 \text{ mol}}$$

   $$= 0.67$$

   The partial pressure of nitrogen = mole fraction of nitrogen × total pressure
   The partial pressure of nitrogen = 0.67×600 torr = 400 torr

3. **2** Heat always flows spontaneously from the object that has the higher temperature to the object with the lower temperature. Therefore, heat travels from X to Y if the temperature of X is higher than the temperature of Y. Choice 2 meets the requirements of spontaneous heat flow because X has a higher temperature than Y.

4. **3** According to Charles' law, the volume of an ideal gas is directly proportional to the Kelvin temperature at constant pressure. When this function is graphically presented, the slope of the function has a positive slope. Choice 3 depicts a positive slope. All other choices display either no change or a negative slope between volume and temperature.

5. **1** By definition, a binary compound is composed of only two different types of elements. The word *binary* means two. Choice 1 (potassium

chloride) is the only compound that is composed of two elements. Ammonium chloride (NH_4Cl) has three elements, potassium chlorate ($KClO_3$) has three elements, and ammonium chlorate (NH_4ClO_3) has four elements.

6. **2** A particle that remains unaffected when it travels through an electric field is neutral in nature (has no charge). All particles with charges (positive or negative) are deflected in the electric field. Only choice 2 (a gamma ray) has no charge. A proton has a positive (+1) charge, an electron has a negative (–1) charge, and an alpha particle has a positive (+2) charge.

7. **4** The major portion of an atom's mass consists of neutrons and protons. A neutron and a proton have an approximate mass of 1 atomic mass unit (amu). The electron is only $\frac{1}{1,836}$ the mass of an atomic mass unit—a minuscule number. Thus, the electron's contribution to an atom's mass is minimal.

8. **1** The atomic number is equal to the number of protons an element has in its nucleus. Therefore, an atom with exactly 15 protons must have an atomic number of 15. According to the periodic table, the element with the atomic number 15 is phosphorus. The number after the element name refers to the mass number, that is, the number of protons plus neutrons (e.g., phosphorus 32 has a total of 32 protons and neutrons).

9. **3** The atomic mass of an element is the weighted average of the naturally occurring isotopes of that element. The weighted average can be calculated by adding the decimal fractions of the isotopes multiplied by their isotopic masses:

Average atomic mass of element $X = \dfrac{72.0 \times 84.9 \text{ amu}}{100} + \dfrac{28.0 \times 87.0 \text{ amu}}{100}$

10. **2** There is a simple rule for understanding nuclear reactions: The sum of superscripts (and subscripts) to the left of the arrow equals the sum of superscripts (and subscripts) to the right of the arrow in a nuclear reaction equation. According to Reference Table J, X must be a beta particle.

11. **4** Hund's rule states that every orbital in a sublevel must be filled with an electron before a second electron can be placed there. Because a p

sublevel has three orbitals, a maximum of six electrons can be placed in it. To have two unpaired electrons, there must be either two or four electrons in the p sublevel. According to electron configurations and the periodic table, groups 14 and 16 have either two or four electrons in the p sublevel. Of the answer choices given, only carbon belongs to group 14. There are no elements in the answer choices that belong to group 16. The electron configuration of carbon is

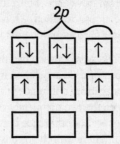

Fluorine has five electrons in the $2p$ sublevel (one unpaired electron), nitrogen has three electrons in the $2p$ sublevel (three unpaired electrons), and beryllium has no electrons in the $2p$ sublevel.

12. 4 The number of neutrons is determined by subtracting the atomic number from the mass number of the element.

Number of neutrons = mass number – atomic number

The correct answer is choice 2 because both atoms contain two neutrons.

$_1^3H = 3 - 1 = 2$ neutrons

$_2^4He = 4 - 2 = 2$ neutrons

13. 1 An empirical formula represents the lowest whole number ratio in which elements make up a compound. Choice 1 (CH_4) is the correct answer because the ratio cannot be reduced to a lower whole number ratio. All other answer choices can be reduced to CH_2 (their empirical formula).

14   3   The potential energy is the stored energy of a compound. The energy is derived from the bonds between the elements that make up the molecule.

15. 4   A balanced equation must follow the law of conservation of mass. According to this law, the number of atoms of each element must be the same on both sides of a reaction equation. Only choice 4 has the same number of hydrogen and oxygen atoms on both sides of the equation.

$$2H_2(g) + O_2(g) \rightarrow 2H_2O(\ell)$$

(4 hydrogens) + (2 oxygens) = (4 hydrogens + 2 oxygens)

16. 4   An atom that has an ionic radius smaller than its atomic radius must be a metal. To have a smaller ionic radius, an element must lose one or more electrons from its valence shell and become a positive ion. Only metals lose electrons (oxidation). Only choice 4 (Rb) is a metal. The rest of the answer choices are nonmetals.

17   3   A polar covalent bond exists between two nonidentical nonmetals in which an electron pair or pairs are shared unequally by the atoms. Usually, the electronegativity difference of approximately 0.4 or greater exists between the two nonmetals. According to the Reference Table K, the electronegative difference between nitrogen (3.1) and hydrogen (2.2) is 0.9. Therefore, choice 3 is the correct answer. All other answer choices depict diatomic molecules. Diatoms form non-polar covalent bonds because the electronegativity difference between atoms is zero.

18. 2   The roman numeral I means that nitrogen has an oxidation number of +1. Nitrogen oxide is composed of nitrogen and oxygen. We know that oxygen's oxidation number is –2. Because nitrogen (I) oxide has an overall molecular charge of zero, the total charge of the nitrogen must be +2 to neutralize oxygen's –2 charge. Because nitrogen is +1, a minimum of two nitrogen atoms are needed to generate an overall positive +2 charge ($2 \times +1 = +2$). Therefore, the correct answer is N_2O (choice 2). The criss-cross method can be used to determine the correct formula:

oxidation
number $\overset{\displaystyle\frown}{}$ $\underset{2}{\overset{+1}{N}}\overset{-2}{\underset{1}{\cancel{O}}} = N_2O$

(1 is never
written)

19. **1** Refer to the periodic table to determine which of the given elements has the strongest metallic characteristic. As we move from left to right in a period, the metallic characteristics of elements decrease. On the other hand, as we move from top to bottom within a group, the metallic characteristics of elements increase. Therefore, elements at the bottom and left side of the periodic table have the strongest metallic characteristics. Of the answer choices given, bismuth (Bi) is the lowest in group 15. Therefore, Bi has the strongest metallic characteristic.

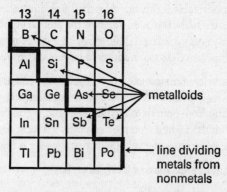

20. **1** The halogens are located in group 17 of the periodic table. The halogen group consists of fluorine, chlorine, bromine, iodine, and astatine. Of the five halogens, at STP fluorine and chlorine are gases, bromine is liquid, and iodine and astatine are solids.

21. **3** In a molecular formula, the number of atoms is indicated by subscripts that are found below and to the right of each of the elements that compose a compound. It is customary not to write the number 1 as a subscript if there is a single atom of a specific element in a compound. For example, the formula $CaCl_2$ indicates that there is one

calcium (Ca) atom and two chlorine (Cl) atoms in the compound. The total number of atoms in the formula $CuSO_4 \cdot 5H_2O$ is 1 Cu + 1 S + 4 O + 5(2 H + 1 O) = 21.

22. 2   Nonmetals and metals combine to form ionic compounds. All ionic compounds have ionic bonds. In an ionic bond, a metal loses electron(s), and a nonmetal gains electron(s). As a result, metals are oxidized and nonmetals are reduced during the formation of ionic bonds. Because metals lose electrons, they acquire positive charges. Conversely, nonmetals acquire negative charges by gaining electrons.

23. 1   Metalloids are elements that display both metallic and nonmetallic characteristics. Metalloids are located immediately on either side of the metal-nonmetal dividing line of the periodic table. Of the answer choices given, only arsenic (As) is located next to the metal-nonmetal dividing line.

24. 2   The periodic table is organized by chemical similarities. According to the periodic table, the mass number, the atomic number, and the principal energy levels are different in a group. However, all elements in a group have the same number of valence electrons.

25. 4   The reactivity of nonmetals follows the same trend as electronegativity. The reactivity increases as we move from left to right within a period and from bottom to top within a group. Therefore, the element with the highest electronegativity is the most reactive nonmetal. Of the answer choices given, the correct answer is fluorine (choice 4).

26. 3   The gram formula mass of $(NH_4)3PO_4$ is

$$N = 14 \text{ g/mol} \times 3 = 42 \text{ g}$$
$$H = 1 \text{ g/mol} \times 12 = 12 \text{ g}$$
$$P = 31 \text{ g/mol} \times 1 = 31 \text{ g}$$
$$O = 16 \text{ g/mol} \times 4 = \underline{64 \text{ g}}$$
$$\text{Total mass} = 149 \text{ g}$$

27. 1   A balanced equation tells us the mole ratio in which compounds react with each other to form products. The coefficients in the balanced equation specify the number of moles of reactants and products in a reaction.

$$CH_4 + 2O_2 \rightarrow CO_2 + 2H_2O$$

According to this balanced equation, for every 1 mole of CH_4 that reacts, 2 moles of O_2 is consumed.

28. 4 Reference Table E lists the solubility of various compounds in water. The compound that is least concentrated has the smallest amount of solute dissolved per given amount of solvent. According to Table E, barium sulfate is nearly insoluble in water. Therefore, barium sulfate is the least concentrated solution.

29. 3 The concentration, or molarity (moles per liter), of a solution is calculated by dividing moles of solute by volume of solution (in liters).

$$\text{Molarity} = \frac{\text{moles of solute}}{\text{volume of solution}}$$

The total number of moles of H_2SO_4 needed to prepare 5 liters of a 2-mole solution is

$$\begin{aligned}\text{Moles of solute} &= \text{molarity} \times \text{volume of solution} \\ &= 2.0 \text{ mol} \times 5.0 \text{ liters} \\ &= 10 \text{ mol}\end{aligned}$$

30. 3 According to the balanced equation, for every 1 mole of Ca(s) that reacts, 1 mole of H_2 gas is produced. Because 40.1 g of Ca(s) is equal to 1 mole of Ca (40.1 g is the molar mass of 1 mole of Ca), 1 mole of H_2 is produced when 4.0 g of Ca is consumed in this reaction. At STP, a mole of gas occupies 22.4 liters of volume. Therefore, the volume of H_2 produced is 22.4 liters.

31. 2 The equilibrium constant (K_{eq}), or equilibrium expression, is defined as the product of the reaction products divided by the product of the reactants. All coefficients in the equation become exponents in the K_{eq} expression.

$4NH_3(g) + 7O_2(g) \rightarrow 4NO_2(g) + 6H_2O(g)$
reactants   products

$$K_{eq} = \frac{[NO_2]^4[H_2O]^6}{[NH_3]^4[O_2]^7}$$

32. 3 A catalyst increases the reaction rate by lowering the activation energy. In the graph, arrow 2 represents the activation energy. Arrow 3 represents the energy of the activated complex. The energy associ-

ated with activated complex activation energy is included within its value. Therefore, when a catalyst lowers activation energy, both arrows 2 and 3 (choice 3) are affected.

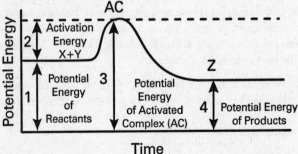

33. 4 The rate of a chemical reaction is dependent on how often molecules (reactants) collide with each other. In effect, any factor that increases the probability of molecules colliding with each other increases the chemical reaction as well. The two factors that can increase the rate of chemical reaction are temperature and concentration of reactants. When temperature is increased, the average kinetic energy of the molecules is also increased. Molecules with higher energy tend to collide with each other more often because they travel at higher velocity. On the other hand, a higher concentration of reactants increases the probability of collisions because there are simply more reactants. In other words, molecules that are in crowded conditions bump into each other more frequently.

34. 1 At equilibrium, the rate of opposing processes must be equal. Remember that equilibrium is dynamic in nature. When a solution displays equilibrium between dissolved and undissolved solute, it means that the rate of dissolving is equal to the rate of crystallization. By definition, at saturation the rate of dissolving equals the rate of crystallization.

35. 4 A solution with a pH greater than 7 must be basic. In this question, the concentration (0.1 M) does not matter. Of the choices given, only KOH is a base. Choice 1 ($C_6H_{12}O_6$) is a sugar and a nonelectrolyte. Therefore, it is a neutral molecule. Choice 2 (CH_3COOH) is an

organic acid with a pH of less than 7. Choice 3 (KCl) is a salt. When in solution, KCl produces a pH of 7.

36. **2** Phenolphthalein, an indicator, turns pink in basic solution. It remains colorless in acidic solution.

37. **4** Reference Table L lists the acid dissociation constants (K_a) of acids (in order of decreasing strength). The K_a value tells us how strong or weak an acid is. The higher the K_a value, the stronger the acid. Therefore, based on Table L, HNO_3 is the strongest acid because it has the highest K_a value of the choices given.

38. **3** Acids can be neutralized by bases. When neutralization occurs, concentrations of $[H^+]$ and $[OH^-]$ are equal. In this question, HCl is an acid and NaOH is a base. Therefore, when HCl is neutralized by NaOH, the hydrogen concentration in the resulting mixture is equal to the concentration of the hydroxide ions.

39. **4** Use the following formula to solve this problem:

$$\text{Volume}_{(acid)} \times \text{molarity}_{(acid)} = \text{volume}_{(base)} \times \text{molarity}_{(base)}$$

$$\text{Molarity}_{(acid)} = \frac{\text{volume}_{(base)} \times \text{molarity}_{(base)}}{\text{volume}_{(acid)}}$$

$$= (20 \text{ ml})(4.0 \text{ M})/20 \text{ ml}$$

$$= 4.0 \text{ M}$$

40. **1** The ionization constant (K_w) is an equilibrium constant, which is affected only by temperature.

41. **2** Amphoteric substances are species that can act as an acid and as a base. In other words, they can donate and receive a proton. According to Reference Table L, of the answer choices given, only NH_3 is an amphoteric species. NH_3 appears on both the left and right sides of the equation (or in both the acid and base columns of Table L).

$NH_3 \rightarrow H^+ + NH_2^-$ (NH_3 acts as an acid here)
$NH_4^+ \rightarrow H^+ + NH_3$ (NH_3 acts as a base here)

42. **3** By definition, a redox reaction is composed of oxidation and reduction reactions occurring at the same time. In fact, *redox* is short for reduction-oxidation.

43. **1** Determine which atom undergoes an oxidation reaction and which one undergoes a reduction reaction.

$$Mg \rightarrow Mg^{2+} + 2e^- \text{ (oxidation)}$$
$$Cr^{3+} + 3e^- \rightarrow Cr \text{ (reduction)}$$

Next, balance the total number of electrons lost and gained.

$$3(Mg \rightarrow Mg^{2+} + 2e^-) = 6e^-$$
$$2(Cr^{3+} + 3e^- \rightarrow Cr) = 6e^-$$

Finally, balance the equation.

$$3Mg + 2Cr^{3+} \rightarrow 3Mg^{2+} + 2Cr$$

Therefore, the sum of the coefficients is $3 + 2 + 3 + 2 = 10$.

44. **2** The oxidation number of oxygen is -2 in NO_2. In choice 1, oxygen has no charge (a free element). In choice 3, oxygen has an oxidation number of -1 (a peroxide). In choice 4, oxygen has a charge of $+2$ (combined with fluorine). Keep in mind that oxygen almost always has a charge of -2 in the combined state. Choices 3 and 4 are rare exceptions.

45. **4** An electrochemical cell, or battery, requires a continuous flow of electrons. To maintain a continuous flow of electrons, a complete circuit must be present. A salt bridge is necessary to form a complete circuit. A salt bridge allows ions to migrate from one half-cell to the other. It also prevents the mixing of the half-cell solutions (refer to the diagram).

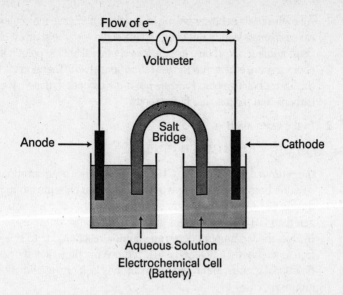

Flow of e⁻

Voltmeter

Anode

Salt
Bridge

Cathode

Aqueous Solution
Electrochemical Cell
(Battery)

46. **4** When a substance is oxidized in a redox reaction it loses electrons. A substance that undergoes oxidation is also called a *reducing agent*.

47. **3** In a reaction, the oxidizing agent itself undergoes reduction. In other words, the oxidizing agent gains electrons. Therefore, its oxidation number is lowered. In the equation given, Ag^+ acquires an electron to become Ag. Because Ag^+ was reduced, it must be the oxidizing agent.

48. **1** Compounds that contain carbon are defined as organic molecules. The study of organic molecules is called *organic chemistry*.

49. **3** Fermentation is part of anaerobic respiration. During fermentation, sugar is oxidized to produce ATP, CO_2, and ethanol (alcohol):

$$\text{glucose} \xrightarrow{\text{zymase}} \text{carbon dioxide} + \text{ethanol}$$

Glucose and zymase are necessary to carry out fermentation, but they are not products of fermentation.

50. 1 All of the answer choices belong to the class of organic molecules called *alkanes*. Within this class, the molecule with the shortest chain (least number of carbons) has the lowest normal boiling point. Of the choices given, ethane has the fewest carbons (two). Therefore, it has the lowest boiling point. Propane has three carbons, butane has four carbons, and pentane has five carbons.

51. 2 In the given reaction

$$CH_3CH_3 + Cl_2 \rightarrow CH_3CH_2Cl + HCl,$$

the saturated hydrocarbon (CH_3CH_3) undergoes a substitution reaction because it exchanges a hydrogen for a chlorine atom, producing CH_3CH_2Cl (notice that this molecule has one fewer hydrogen than CH_3CH_3). Choice 1 is incorrect because saturated hydrocarbons cannot undergo an addition reaction due to lack of free, open bonds. Choices 3 and 4 require the presence of organic functional groups. Saturated hydrocarbons lack organic functional groups.

52. 1 Saturated hydrocarbons are composed of single bonds only. Of the answer choices given, only ethane is made of single bonds. Ethane belongs to the class of organic molecules known as *alkanes*. The primary characteristic of alkanes is that they only contain single bonds. Alkanes are easily identified by name: They all end in -*ane*.

53. 1 According to Reference Table K, the electronegativity of elements decreases as we move down group 16.

54. 2 According to Boyle's law, the volume and the pressure of gases are inversely proportional to each other. As one increases, the other decreases. Therefore, if pressure decreases from 760 to 380 torr, the volume of the gas increases.

55. 1 Entropy, or measurement of randomness, increases as a substance changes its phase from solid to liquid to gas. Conversely, entropy decreases as a substance transforms itself from a gas to a liquid to a solid. In this problem, carbon changes from a gaseous phase to a solid phase. Therefore, entropy must have decreased.

56. **2** The rate of a chemical reaction is directly proportional to the number of effective collisions between molecules. In other words, the higher the number of collisions, the greater the reaction rate. Larger surface area increases the probability of molecules colliding with each other. Consequently, the reaction rate increases.

PART 2
Group 1—Matter and Energy

57. **4** Use the following formula to solve this problem:

$\Delta H = mc\Delta T$

ΔH = total number of calories

m = mass of water

c = specific heat of water

ΔT = change in temperature

$\Delta H = (15 \text{ g})(1 \text{ cal/g} \cdot {}^{\circ}\text{C})(10{}^{\circ}\text{C})$

$\quad = 150 \text{ cal}$

(Hint: c has a numerical value of 1 and, therefore, does not affect the answer.)

58. **4** The liquid and the solid phases are present when the liquid is freezing to become a solid. During the phase change, the temperature remains constant. Therefore, line *DE* must be the curve that represents the time when both the liquid and the solid phases are present.

59. **3** An exothermic reaction releases energy. During phase change, energy is released when a substance changes from a gas to a liquid to a solid. Therefore, the correct answer is choice 3 because it depicts a liquid changing to a solid.

$H_2O(\ell) \rightarrow H_2O(s)$

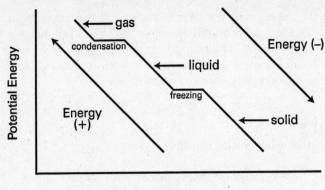

— Energy is released as a substance changes from gaseous to liquid to solid phase.

— Energy is absorbed as a substance changes from solid to liquid to gaseous phase.

60. **4** The kinetic theory of gas states the following:

1. The mass of the gas particle is minuscule compared to the volume it occupies.
2. Gas molecules randomly travel in a straight line.
3. There are no attractive or repulsive forces between gas molecules.
4. Energy is transferred between colliding gas molecules, but energy is not lost.

Based on this list, the correct answer is choice 4. All other answer choices contradict the kinetic theory of gases.

61. **1** By definition, a compound is a substance that is composed of two or more elements with a fixed mass. Such composition is termed *homogenous composition*. On the other hand, a mixture consists of two or more substances that have variable compositions and can be homogenous or heterogeneous in nature.

Group 2—Atomic Structure

62. **1** The atomic number of an element is equal to the number of protons in its nucleus. Because the ion in question has five protons, its atomic number must be five as well.

63. **2** Electrons can move between energy levels within an atom. As energy levels move farther and farther away from the nucleus, more energy is associated with them. In other words, the first energy level has the least amount of energy related to it. Therefore, if an electron wants to move to a higher energy level, it must absorb energy. Electrons that jump to a higher energy level are referred to as *excited electrons*. In the same manner, when electrons return to a lower energy level from a higher level, they emit energy. Based on the behavior of migrating electrons, choice 2 is the correct answer.

64. **4** Use the periodic table to answer this question. A period of the table tells us the principal quantum number of the highest energy level of an atom in that period. Therefore, zinc's highest principal quantum number is 4 because it is found in period 4.

65. **2** Use Reference Table K to answer this question. A first ionization energy of 176 kcal corresponds to magnesium (Mg), which is located in group 2 of the periodic table. All elements in group 2 contain two valence electrons.

66. **4** Nickel has an atomic number of 28 and is found in period 4 of the periodic table. Nickel's electron configuration is $1s^2 2s^2 2p^6 3s^2 3p^6 3d^8 4s^2$. Based on the electron configuration, we can conclude that the total number of occupied s orbitals in an atom of nickel in the ground state is four (choice 4).

Group 3—Bonding

67. **2** Nickel (II) has a charge of +2. Hypochlorite (ClO^-) has a charge of -1. Therefore, the chemical formula for nickel (II) hypochlorite is $Ni(ClO)_2$. Use the criss-cross method:

$$Ni^{+2}(ClO)^{-1}_{} = Ni(ClO)_2$$

68. 2 The most stable compound has the greatest negative Gibb's free energy charge ($-\Delta G$). The right column of Reference Table G provides ΔG values. The greatest $-\Delta G$ occurs with the formation of $CO_2(g)$. The middle column is the ΔH column. It is used to determine which compound releases (exothermic) or absorbs (endothermic) the most energy on forming.

69. 4 Krypton is a noble gas. All noble gases are nonpolar in nature. Therefore, of the four choices given, only van der Waals forces can exist between krypton molecules. The ionic and molecule-ion forces require charged or polar particles. A covalent bond is an intramolecular bond and not an intermolecular bond.

70. 2 Hydrogen bonds are intermolecular bonds that occur between molecules that contain hydrogen and atoms of another element of higher electronegativity. The attraction of a hydrogen atom from one molecule and an atom of higher electronegativity from an adjacent molecule constitutes a hydrogen bond. Hydrogen bonds are exceptionally strong between molecules of hydrogen fluoride (HF) and between molecules of water (H_2O). For that reason, these substances have unusually high boiling points.

71. 3 Electronegative values tell us an atom's affinity for electrons. The higher the electronegativity, the greater the attraction for electrons. According to Reference Table K, nitrogen has the highest electronegativity of the four elements given in the answer choices.

Group 4—Periodic Table

72. 2 Use the periodic table and Reference Table P to answer this question. Table P indicates the atomic radius of elements. Find the elements that correspond to electron configurations given in each answer choice. According to Table P, $1s^2 2s^1$ (Li) has the largest covalent radius.

73. 1 You must recall the various characteristics of elements that are found in the periodic table to answer this question. Recall that ions of transition elements are usually colored, both as solids and in aqueous solution. Of the four answer choices given, only Cu^{2+} represents a transition element ion.

74. 1 All of the choices are metals. The most active metals are in group 1 (alkali metals). The next most active metals are in group 2 (alkaline earth metals). Also, within a group of metals, activity increases as one goes down the list. Calcium (Ca) is not as active as strontium (Sr) because it is in the same group but above it. Potassium (K) is more active than strontium because it is in group 1. Iron and copper are transition elements, which are not nearly as active as alkali metals or alkaline earth metals.

75. 4 Metals are better conductors than nonmetals. In the answer choices given, only choice 4 (silver) is a metal. All other choices are non-metals.

76. 2 The compound XO is composed of the metal X and oxygen. Oxygen has a charge oxidation number of –2. Therefore, in the compound XO, X must have a +2 charge. All elements that belong to group 2 have a charge of +2.

Group 5—Mathematics of Chemistry

77. 1 According to Graham's law, the rate at which a gas diffuses is proportional to its mass. The lighter the gas molecule is, the faster its diffusion rate. Conversely, gases that are heavier in mass diffuse at a lower rate. Of the choices given, He (helium) has the lowest molecular mass. Therefore, it diffuses most rapidly.

78. 4 One mole of hydrogen (H_2) contains 6×10^{23} molecules. Therefore, in 0.25 mole (¼ of 1 mole), there are $0.25(6.0 \times 10^{23}$ molecules) = 1.5×10^{23} molecules.

79. 3 According to Boyle's law, the volume of a gas is inversely proportional to the pressure. On the other hand, Charles' law states that the volume of a gas is directly proportional to the temperature (in kelvin). Therefore, a 1.00-mole sample of an ideal gas decreases in volume when the pressure increases (inverse relationship) and the temperature decreases (direct relationship).

80. 4 According to the laws of freezing-point depression, when solutes are added to water, the freezing point is lowered. One factor that determines how much the freezing point is lowered (depressed) is the number of particles (solute) present in the water. The greater the

number of dissolved particles, the more the freezing point is depressed. When dissolved in water, ionic compounds produce more particles than organic compounds. Of the four choices given, only choice 4 (NaOH) is an ionic compound. Therefore, an aqueous solution of NaOH has the lowest freezing point.

81 1 Start solving this problem by assuming that there is a 100-g sample of the compound. Because the compound is 85% Ag and 15% F, assume that there are 85 g of Ag and 15 g of F in the sample. Next, convert 85 g of Ag and 15 g of F to moles by dividing them by their respective atomic masses.

$$Ag = \frac{85\ g}{108\ g} = 0.787\ mol$$

$$F = \frac{15\ g}{19\ g} = 0.789\ mol$$

Divide both calculated moles by the lower of the two to obtain the smallest whole number ratio possible.

Ag = 0.787 mol/0.787 mol = 1
F = 0.789 mol/0.787 mol = 1

These values then become subscripts in the empirical formula. Based on the mole ratio, the empirical formula must be AgF. (Remember, the number 1 is never written.)

Group 6—Kinetics and Equilibrium

82. 1 Reference Table M lists the solubility products (K_{sp}) of various compounds. The solubility product states a compound's ability to dissolve. The higher the solubility product, the more soluble a compound is in a solution. Conversely, the lower the solubility product, the less soluble a compound is in a solution. According to Table M, only AgI (choice 1) has a lower K_{sp} than $PbCO_3$. Therefore, AgI must be less soluble than $PbCO_3$.

83. 1 The reaction stated in the question is an endothermic reaction. According to the reaction, heat is consumed when HBr is decomposed to H_2 and Br_2. If temperature is increased, more heat is added to the reaction. When heat is added to an endothermic reaction, reac-

tants rapidly react to form more products. In this case, the reaction shifts to the right, causing additional HBr to decompose to H_2 and Br_2. As a result, the concentration of products increases whereas the concentration of HBr decreases. General rule of the Le Châtelier principle: An increase in temperature causes a shift in the direction that absorbs heat (endothermic reactions). A decrease in temperature causes a shift in the direction that releases heat (exothermic reaction).

$$\downarrow 2 \text{ HBr(g)} + \uparrow 17 \text{ kcal} \rightleftharpoons \uparrow H_2(g) + \uparrow Br_2 \text{ (g)}$$

② decrease in concentration of reactants (as a result of shift)

① increased temperature causes a shift to the *right*

② increase in the concentration of products (as a result of shift)

84. 4 A reaction is said to have reached a state of *dynamic equilibrium* when both the forward and reverse reactions proceed at equal rates. In other words, at equilibrium both the reactants and products form at the same rate. As a result, there is no net gain or loss in the concentrations of reactants and products at equilibrium. The concentration of reactants and products need not be equal but must remain constant.

85. 3 All reactions that are spontaneous in nature have negative free energies of formation (ΔG°_f). ΔG°_f of various reactions are listed in Reference Table G. Of the four answer choices given, only the reaction in choice 3 has a negative ΔG°_f value. Therefore, it must be a spontaneous reaction.

86. 2 The reaction $A + B \rightarrow C$ + energy is an exothermic reaction. Whenever the energy is on the right side of the arrow (on the side of the products), the reaction is exothermic. In an exothermic reaction, the products have less potential energy than the reactants. Choice 2 shows that the product (C) has a lower potential energy than the reactants ($A + B$).

Group 7—Acids and Bases

87. 4 Potassium chloride (KCl) is a salt. Salt is one of the products in a neutralization reaction. A generic neutralization looks like this:

HA + BOH → BA + HOH
acid   base   salt water

The specific neutralization for the formation of KCl looks like this:

HCl + KOH → KCl + HOH

Based on Reference Table L, HCl is a strong acid. Strong bases are formed between alkali metals (Li, Na, K) and the hydroxide group (OH⁻). Therefore, KOH is a strong base.

88.   2   By definition, a Brönsted-Lowry acid is able to donate a proton (H^+), whereas a Brönsted-Lowry base is capable of accepting a proton. In other words, in Brönsted-Lowry acid-base pairs, the acid differs from the base by a proton. In the given reaction, there are two Brönsted-Lowry acid-base pairs. The first pair is HSO_4^- and SO_4^{-2}. The second pair is H_2O and H_3O^+. Only the first pair, HSO_4^- and SO_4^{-2}, are given in the answer choices. (Hint: In a Brönsted-Lowry acid-base pair, the acid contains one more H^+ than the base.)

89.   3   The measurement of protons or H^+ ions in solutions is called the *pH*. The mathematical formula that represents pH is $-\log[H^+]$. Because the aqueous solution in the question has a hydrogen concentration of 1×10^{-8} mole per liter, the pH of the solution is equal to pH = $-\log[1 \times 10^{-8}] = 8$. Any solution that has a pH greater than 7 is called a *basic solution*.

90.   4   The ionization constant of water is 1×10^{-14}.

$$K_w = [OH^-][H_3O^+] = 1.0 \times 10^{-14}$$

Therefore, regardless of the $[OH^-]$, the $[OH^-][H_3O^+]$ product is 1×10^{-14}.

91.   1   In the given reaction

$$KOH + HNO_3 \rightarrow KNO_3 + H_2O,$$

KOH is a base, HNO_3 is an acid, KNO_3 is a salt, and H_2O is water. When an acid and a base react to form a salt and water, it is called a *neutralization reaction* (choice 1). In an esterification reaction, a carboxylic acid reacts with an alcohol to produce an organic compound, called an *ester*, and water; in a substitution reaction, an atom or group of atoms is replaced by another atom or group of atoms; and in an addition reaction, an atom or group of atoms is added to an unsaturated bond (double or triple bond) of an existing molecule.

Group 8—Redox and Electrochemistry

92. 4   In a balanced equation, all of the atoms of each element on both sides of the equation must be equal. The quickest way to balance any given equation is by using the trial-and-error method. The reaction in question is balanced as follows:

$$MnO_2 + 4HCl \rightarrow MnCl_2 + 2H_2O + Cl_2$$

Based on this balanced equation, the coefficient of HCl is 4 (choice 4).

93. 4   Reference Table N lists the standard reduction potential of half-cells from the highest to the lowest value. The standard reduction potential of a hydrogen cell is 0.00 V. Of the four answer choices given, only choice 4 ($Pb^{2+} + 2e^- \rightarrow Pb$) has a reduction potential (–0.13 V) lower than that of hydrogen.

94. 2   In a spontaneous redox reaction, the species that is being reduced has a more positive reduction potential than the species that is being oxidized. In other words, the species that is being reduced must appear above the species being oxidized on Reference Table N. In choice 2, Au^{3+} is reduced to Au and Al is oxidized to Al^{3+}. In Table N, Au^{3+} has a higher reduction potential than Al^{3+}. That means that the reaction in choice 2 is a spontaneous reaction.

95. 1   Determine the species that undergoes reduction and the species that undergoes oxidation. In the given reaction, Zn^{2+} undergoes reduction and Mg undergoes oxidation.

$$Zn^{2+} + 2e^- \rightarrow Zn$$
$$Mg \rightarrow Mg^{2+} + 2e^-$$

Next, determine the electrode potentials of the Zn and Mg from Reference Table N.

$$Zn = -0.76 \text{ V}$$
$$Mg = -2.37 \text{ V}$$

Finally, find the cell voltage (E^0) by subtracting the electrode potential of the species being oxidized from that of the species being reduced.

$$E^0 = E^0_{reduced} - E^0_{oxidized}$$
$$= (-0.76 \text{ V}) - (-2.37 \text{ V}) = +1.61 \text{ V}$$

96. 1 In an electrochemical cell or battery, electrons flow from the electrode where oxidation occurs to the electrode where reduction takes place. In the given figure, oxidation occurs at Al and reduction occurs at Ni. Therefore, electrons flow from Al(s) to Ni(s).

Group 9—Organic Chemistry

97. 2 C_4H_{10} is also known as *butane*. Butane belongs to a class of molecules called the *alkanes*. All alkanes have the general formula C_nH_{2n+2}.

98. 1 Structural isomers have the same molecular formulas but different structural formulas, which means that the isomers must have the same number of atoms of each element. In the problem, the compound in question is composed of two carbons, six hydrogens, and one oxygen. The only choice that has the same number of atoms of each element, but a different structural formula, is choice 1. Choice 2 is exactly the same as the compound mentioned in the question and, thus, is not an isomer.

99. 3 By definition, tertiary alcohols have no hydrogen atoms attached to the carbon that is bonded to the —OH functional group. Therefore, the carbon atom that has the —OH functional group must be bonded to three other carbon atoms. Recall that a carbon can form up to four single bonds.

100. 2 Proteins are polymers of amino acid monomers. All other choices are monomers. None of them can be broken down further into simpler, repeating units.

101. 3 Glycerol contains three carbon atoms. A glycerol molecule is a trihydroxy alcohol. Each alcohol accounts for each one of the carbons. The prefix *tri-* means three. Therefore, glycerol must contain three carbon atoms.

Group 10—Applications of Chemical Principles

102. 1   The definition of *corrosion* is a metal undergoing oxidation to become an ion. Therefore, corrosion must be an oxidation-reduction reaction.

103. 3   Petroleum is a mixture of numerous varieties of organic compounds. Each one of these compounds has a unique boiling point. Based on differences in boiling points, various compounds in mixtures can be separated. The process that separates compounds in a mixture based on boiling point difference is called *fractional distillation.*

104. 3   An automobile battery carries out the following redox reaction:

$$Pb(s) + PbO_2(s) + 2H_2SO_4 \rightarrow 2PbSO_4(s) + 2H_2O(\ell)$$

According to this reaction, H_2SO_4 is the electrolyte in an automobile battery.

105. 2   A battery consists of two electrodes where reduction and oxidation (redox) reactions take place spontaneously, converting chemical energy into electrical energy. This, by definition, is an electrochemical cell. A redox reaction is a chemical process in which one species loses electrons (oxidation) whereas the second species gains electrons (reduction).

106. 3   The elements that can be found in nature in the free state are the ones that are least reactive. Of the four answer choices given, Au (gold) is the least reactive. Ca, Ba, and Al are reactive metals. In addition, Au is found near the top of Reference Table N, indicating that Au prefers to undergo reduction and remain uncombined. Therefore, Au is likely to be found in nature in its free state.

Group 11—Nuclear Chemistry

107. 1   To be considered for use in geological dating, a radioisotope must have a long half-life. Consult Reference Table H to determine the half-lives of the isotopes given in the answer choices. According to Table H, uranium 238 has the longest half-life.

108. 1   In a fusion reaction, two nuclei combine to form a more massive single nucleus. The only choice that correctly depicts fusion is choice 1, in which ^3_1H and ^1_1H combine to form ^4_2He. Choice 2 depicts artificial transmutation, whereas choice 3 is an example of beta decay and choice 4 represents alpha decay. Generally, nuclear fusion involves the conversion of hydrogen (H) to helium (He).

109. 4 A coolant is used in a nuclear reactor to stabilize the temperature at a reasonable level. The only logical answer is choice 4 (heavy water).

110. 3 ^{233}U is an isotope of ^{235}U. Isotopes have the same number of protons (atomic number) but a different number of neutrons (mass number). Isotopes also have similar chemical properties, because they have the same configuration of electrons.

111. 1 In nuclear reactors, control rods serve to control the rate of fission by absorbing neutrons. Therefore, any element that has the ability to absorb neutrons can serve as a control rod. Boron and cadmium must be able to absorb neutrons because they serve as control rods.

Group 12—Laboratory Activities

112. 1 To measure the effect of surface area, two trials must be picked in which all factors remain constant except the particle size of $NaHCO_3$. In other words, concentration and temperature of HCl must remain the same in both trials. After close inspection of the table, it is clear that trials A and B or trials D and E can be used. Because trials D and E do not appear in one of the answer choices, choice 1 (trials A and B) is the only logical answer.

113. 4 The rate of reaction, or how fast a reaction takes place, depends on how frequently molecules collide with each other. Therefore, any factors that increase the probability of molecules colliding with each other also increase the reaction rate. The rate of a chemical reaction increases with an increase in concentrations of reactants, temperature, and surface area (smaller particle size). Therefore, trial D (choice 4) is the correct answer.

114. 2 Percent error can be calculated by using the following formula:

$$\% \text{ Error} = \frac{\text{experimental value} - \text{accepted value}}{\text{accepted value}} \times 100$$

$$= \frac{88 \text{ cal/g} - 80 \text{ cal/g}}{80 \text{ cal/g}} \times 100 = 10\%$$

115. 2 The number 52.6 cm has three significant figures. The number 1.214 cm has four significant figures. Therefore, the product of the two numbers must have three significant figures.

$$52.6 \text{ cm} \times 1.214 \text{ cm} = 63.9 \text{ cm}^2$$

116. **3** Examine the two rulers closely. From your observation, you should realize that ruler *A* is marked every centimeter. On the other hand, ruler *B* is marked every tenth of a centimeter. The common practice is to carry precision of measurement one order of magnitude "finer" than the calibrations on the instrument. Therefore, when using ruler *A*, precise measurement can be estimated to the nearest tenth, whereas ruler *B* can precisely measure to the nearest hundredth.

ABOUT THE AUTHOR

Nilanjan Sen is an honors graduate of the University of Rochester and the New York University Graduate School. Currently, he holds the position of adjunct professor of biology at LaGuardia College. Previously, he served as assistant professor of biology at Westchester College. He is a former instructor of MCAT at The Princeton Review. Nilanjan is also the owner and chief executive officer of Indus Publishing Corporation and SGS Worldwide Media, with offices in Wayland, New York; London; and Singapore.

The University of the State of New York

REGENTS HIGH SCHOOL EXAMINATION

CHEMISTRY

ANSWER SHEET

☐ Male

Student .... Sex: ☐ Female

Teacher ....

School ....

Record all of your answers on this answer sheet in accordance with the instructions on the front cover of the test booklet.

Part I (65 credits)

1	1 2 3 4	21	1 2 3 4	41	1 2 3 4
2	1 2 3 4	22	1 2 3 4	42	1 2 3 4
3	1 2 3 4	23	1 2 3 4	43	1 2 3 4
4	1 2 3 4	24	1 2 3 4	44	1 2 3 4
5	1 2 3 4	25	1 2 3 4	45	1 2 3 4
6	1 2 3 4	26	1 2 3 4	46	1 2 3 4
7	1 2 3 4	27	1 2 3 4	47	1 2 3 4
8	1 2 3 4	28	1 2 3 4	48	1 2 3 4
9	1 2 3 4	29	1 2 3 4	49	1 2 3 4
10	1 2 3 4	30	1 2 3 4	50	1 2 3 4
11	1 2 3 4	31	1 2 3 4	51	1 2 3 4
12	1 2 3 4	32	1 2 3 4	52	1 2 3 4
13	1 2 3 4	33	1 2 3 4	53	1 2 3
14	1 2 3 4	34	1 2 3 4	54	1 2 3
15	1 2 3 4	35	1 2 3 4	55	1 2 3
16	1 2 3 4	36	1 2 3 4	56	1 2 3
17	1 2 3 4	37	1 2 3 4		
18	1 2 3 4	38	1 2 3 4		
19	1 2 3 4	39	1 2 3 4		
20	1 2 3 4	40	1 2 3 4		

Your answers for Part II should be placed in the proper spaces on the back of this sheet.

Part II (35 credits)

Answer the questions in only seven of the twelve groups in this part. Be sure to mark the answers to the groups of questions you choose in accordance with the instructions on the front cover of the test booklet. Leave blank the five groups of questions you do not choose to answer.

Group 1 Matter and Energy				
57	1	2	3	4
58	1	2	3	4
59	1	2	3	4
60	1	2	3	4
61	1	2	3	4

Group 2 Atomic Structure				
62	1	2	3	4
63	1	2	3	4
64	1	2	3	4
65	1	2	3	4
66	1	2	3	4

Group 3 Bonding				
67	1	2	3	4
68	1	2	3	4
69	1	2	3	4
70	1	2	3	4
71	1	2	3	4

Group 4 Periodic Table				
72	1	2	3	4
73	1	2	3	4
74	1	2	3	4
75	1	2	3	4
76	1	2	3	4

Group 5 Mathematics of Chemistry				
77	1	2	3	4
78	1	2	3	4
79	1	2	3	4
80	1	2	3	4
81	1	2	3	4

Group 6 Kinetics and Equilibrium				
82	1	2	3	4
83	1	2	3	4
84	1	2	3	4
85	1	2	3	4
86	1	2	3	4

Group 7 Acids and Bases				
87	1	2	3	4
88	1	2	3	4
89	1	2	3	4
90	1	2	3	4
91	1	2	3	4

Group 8 Redox and Electrochemistry				
92	1	2	3	4
93	1	2	3	4
94	1	2	3	4
95	1	2	3	4
96	1	2	3	4

Group 9 Organic Chemistry				
97	1	2	3	4
98	1	2	3	4
99	1	2	3	4
100	1	2	3	4
101	1	2	3	4

Group 10 Applications of Chemical Principles				
102	1	2	3	4
103	1	2	3	4
104	1	2	3	4
105	1	2	3	4
106	1	2	3	4

Group 11 Nuclear Chemistry				
107	1	2	3	4
108	1	2	3	4
109	1	2	3	4
110	1	2	3	4
111	1	2	3	4

Group 12 Laboratory Activities				
112	1	2	3	4
113	1	2	3	4
114	1	2	3	4
115	1	2	3	4
116	1	2	3	4

I do hereby affirm, at the close of this examination, that I had no unlawful knowledge of the questions or answers prior to the examination and that I have neither given nor received assistance in answering any of the questions during the examination.

_____

Signature

[OVER]

The University of the State of New York

REGENTS HIGH SCHOOL EXAMINATION

CHEMISTRY

ANSWER SHEET

☒ Male

Student .... Sex: ☐ Female

Teacher ....

School ....

Record all of your answers on this answer sheet in accordance with the instructions on the front cover of the test booklet.

Part I (65 credits)

1	1 2 3 4	21	1 2 3 4	41	1 2 3 4
2	1 2 3 4	22	1 2 3 4	42	1 2 3 4
3	1 2 3 4	23	1 2 3 4	43	1 2 3 4
4	1 2 3 4	24	1 2 3 4	44	1 2 3 4
5	1 2 3 4	25	1 2 3 4	45	1 2 3 4
6	1 2 3 4	26	1 2 3 4	46	1 2 3 4
7	1 2 3 4	27	1 2 3 4	47	1 2 3 4
8	1 2 3 4	28	1 2 3 4	48	1 2 3 4
9	1 2 3 4	29	1 2 3 4	49	1 2 3 4
10	1 2 3 4	30	1 2 3 4	50	1 2 3 4
11	1 2 3 4	31	1 2 3 4	51	1 2 3 4
12	1 2 3 4	32	1 2 3 4	52	1 2 3 4
13	1 2 3 4	33	1 2 3 4	53	1 2 3
14	1 2 3 4	34	1 2 3 4	54	1 2 3
15	1 2 3 4	35	1 2 3 4	55	1 2 3
16	1 2 3 4	36	1 2 3 4	56	1 2 3
17	1 2 3 4	37	1 2 3 4		
18	1 2 3 4	38	1 2 3 4		
19	1 2 3 4	39	1 2 3 4		
20	1 2 3 4	40	1 2 3 4		

Your answers for Part II should be placed in the proper spaces on the back of this sheet.

FOR TEACHER USE ONLY

Credits

Part I ....
(Use table below)

Part II ....

Total ....

Rater's Initials: ....

Part I Credits

Directions to Teacher:

In the table below, draw a circle around the number of right answers and the adjacent number of credits. Then write the number of credits (not the number right) in the space provided above.

No. Right	Credits		No. Right	Credits
56	65		28	40
55	64		27	39
54	63		26	38
53	62		25	38
52	61		24	37
51	61		23	36
50	60		22	35
49	59		21	34
48	58		20	33
47	57		19	32
46	56		18	31
45	55		17	30
44	54		16	30
43	53		15	29
42	53		14	28
41	52		13	26
40	51		12	24
39	50		11	22
38	49		10	20
37	48		9	18
36	47		8	16
35	46		7	14
34	46		6	12
33	45		5	10
32	44		4	8
31	43		3	6
30	42		2	4
29	41		1	2
			0	0

No. right ....

Part II (35 credits)

Answer the questions in only seven of the twelve groups in this part. Be sure to mark the answers to the groups of questions you choose in accordance with the instructions on the front cover of the test booklet. Leave blank the five groups of questions you do not choose to answer.

Group 1
Matter and Energy

57	1	2	3	4
58	1	2	3	4
59	1	2	3	4
60	1	2	3	4
61	1	2	3	4

Group 2
Atomic Structure

62	1	2	3	4
63	1	2	3	4
64	1	2	3	4
65	1	2	3	4
66	1	2	3	4

Group 3
Bonding

67	1	2	3	4
68	1	2	3	4
69	1	2	3	4
70	1	2	3	4
71	1	2	3	4

Group 4
Periodic Table

72	1	2	3	4
73	1	2	3	4
74	1	2	3	4
75	1	2	3	4
76	1	2	3	4

Group 5
Mathematics of Chemistry

77	1	2	3	4
78	1	2	3	4
79	1	2	3	4
80	1	2	3	4
81	1	2	3	4

Group 6
Kinetics and Equilibrium

82	1	2	3	4
83	1	2	3	4
84	1	2	3	4
85	1	2	3	4
86	1	2	3	4

Group 7
Acids and Bases

87	1	2	3	4
88	1	2	3	4
89	1	2	3	4
90	1	2	3	4
91	1	2	3	4

Group 8
Redox and Electrochemistry

92	1	2	3	4
93	1	2	3	4
94	1	2	3	4
95	1	2	3	4
96	1	2	3	4

Group 9
Organic Chemistry

97	1	2	3	4
98	1	2	3	4
99	1	2	3	4
100	1	2	3	4
101	1	2	3	4

Group 10
Applications of Chemical Principles

102	1	2	3	4
103	1	2	3	4
104	1	2	3	4
105	1	2	3	4
106	1	2	3	4

Group 11
Nuclear Chemistry

107	1	2	3	4
108	1	2	3	4
109	1	2	3	4
110	1	2	3	4
111	1	2	3	4

Group 12
Laboratory Activities

112	1	2	3	4
113	1	2	3	4
114	1	2	3	4
115	1	2	3	4
116	1	2	3	4

I do hereby affirm, at the close of this examination, that I had no unlawful knowledge of the questions or answers prior to the examination and that I have neither given nor received assistance in answering any of the questions during the examination.

_____
Signature

[OVER]

The University of the State of New York

REGENTS HIGH SCHOOL EXAMINATION

CHEMISTRY

ANSWER SHEET

Student .... Sex: ☐ Male   ☐ Female

Teacher ....

School ....

Record all of your answers on this answer sheet in accordance with the instructions on the front cover of the test booklet.

Part I (65 credits)

1	1 2 3 4	21	1 2 3 4	41	1 2 3 4
2	1 2 3 4	22	1 2 3 4	42	1 2 3 4
3	1 2 3 4	23	1 2 3 4	43	1 2 3 4
4	1 2 3 4	24	1 2 3 4	44	1 2 3 4
5	1 2 3 4	25	1 2 3 4	45	1 2 3 4
6	1 2 3 4	26	1 2 3 4	46	1 2 3 4
7	1 2 3 4	27	1 2 3 4	47	1 2 3 4
8	1 2 3 4	28	1 2 3 4	48	1 2 3 4
9	1 2 3 4	29	1 2 3 4	49	1 2 3 4
10	1 2 3 4	30	1 2 3 4	50	1 2 3 4
11	1 2 3 4	31	1 2 3 4	51	1 2 3 4
12	1 2 3 4	32	1 2 3 4	52	1 2 3 4
13	1 2 3 4	33	1 2 3 4	53	1 2 3
14	1 2 3 4	34	1 2 3 4	54	1 2 3
15	1 2 3 4	35	1 2 3 4	55	1 2 3
16	1 2 3 4	36	1 2 3 4	56	1 2 3
17	1 2 3 4	37	1 2 3 4		
18	1 2 3 4	38	1 2 3 4		
19	1 2 3 4	39	1 2 3 4		
20	1 2 3 4	40	1 2 3 4		

Your answers for Part II should be placed in the proper spaces on the back of this sheet.

Part II (35 credits)

Answer the questions in only seven of the twelve groups in this part. Be sure to mark the answers to the groups of questions you choose in accordance with the instructions on the front cover of the test booklet. Leave blank the five groups of questions you do not choose to answer.

Group 1 Matter and Energy				
57	1	2	3	4
58	1	2	3	4
59	1	2	3	4
60	1	2	3	4
61	1	2	3	4

Group 2 Atomic Structure				
62	1	2	3	4
63	1	2	3	4
64	1	2	3	4
65	1	2	3	4
66	1	2	3	4

Group 3 Bonding				
67	1	2	3	4
68	1	2	3	4
69	1	2	3	4
70	1	2	3	4
71	1	2	3	4

Group 4 Periodic Table				
72	1	2	3	4
73	1	2	3	4
74	1	2	3	4
75	1	2	3	4
76	1	2	3	4

Group 5 Mathematics of Chemistry				
77	1	2	3	4
78	1	2	3	4
79	1	2	3	4
80	1	2	3	4
81	1	2	3	4

Group 6 Kinetics and Equilibrium				
82	1	2	3	4
83	1	2	3	4
84	1	2	3	4
85	1	2	3	4
86	1	2	3	4

Group 7 Acids and Bases				
87	1	2	3	4
88	1	2	3	4
89	1	2	3	4
90	1	2	3	4
91	1	2	3	4

Group 8 Redox and Electrochemistry				
92	1	2	3	4
93	1	2	3	4
94	1	2	3	4
95	1	2	3	4
96	1	2	3	4

Group 9 Organic Chemistry				
97	1	2	3	4
98	1	2	3	4
99	1	2	3	4
100	1	2	3	4
101	1	2	3	4

Group 10 Applications of Chemical Principles				
102	1	2	3	4
103	1	2	3	4
104	1	2	3	4
105	1	2	3	4
106	1	2	3	4

Group 11 Nuclear Chemistry				
107	1	2	3	4
108	1	2	3	4
109	1	2	3	4
110	1	2	3	4
111	1	2	3	4

Group 12 Laboratory Activities				
112	1	2	3	4
113	1	2	3	4
114	1	2	3	4
115	1	2	3	4
116	1	2	3	4

I do hereby affirm, at the close of this examination, that I had no unlawful knowledge of the questions or answers prior to the examination and that I have neither given nor received assistance in answering any of the questions during the examination.

_____

Signature

[OVER]

The University of the State of New York

REGENTS HIGH SCHOOL EXAMINATION

CHEMISTRY

ANSWER SHEET

☐ Male

Student .... Sex: ☐ Female

Teacher ....

School ....

Record all of your answers on this answer sheet in accordance with the instructions on the front cover of the test booklet.

Part I (65 credits)

1	1 2 3 4		**21**	1 2 3 4		**41**	1 2 3 4
2	1 2 3 4		**22**	1 2 3 4		**42**	1 2 3 4
3	1 2 3 4		**23**	1 2 3 4		**43**	1 2 3 4
4	1 2 3 4		**24**	1 2 3 4		**44**	1 2 3 4
5	1 2 3 4		**25**	1 2 3 4		**45**	1 2 3 4
6	1 2 3 4		**26**	1 2 3 4		**46**	1 2 3 4
7	1 2 3 4		**27**	1 2 3 4		**47**	1 2 3 4
8	1 2 3 4		**28**	1 2 3 4		**48**	1 2 3 4
9	1 2 3 4		**29**	1 2 3 4		**49**	1 2 3 4
10	1 2 3 4		**30**	1 2 3 4		**50**	1 2 3 4
11	1 2 3 4		**31**	1 2 3 4		**51**	1 2 3 4
12	1 2 3 4		**32**	1 2 3 4		**52**	1 2 3 4
13	1 2 3 4		**33**	1 2 3 4		**53**	1 2 3
14	1 2 3 4		**34**	1 2 3 4		**54**	1 2 3
15	1 2 3 4		**35**	1 2 3 4		**55**	1 2 3
16	1 2 3 4		**36**	1 2 3 4		**56**	1 2 3
17	1 2 3 4		**37**	1 2 3 4			
18	1 2 3 4		**38**	1 2 3 4			
19	1 2 3 4		**39**	1 2 3 4			
20	1 2 3 4		**40**	1 2 3 4			

No. right...............

Your answers for Part II should be placed in the proper spaces on the back of this sheet.

Part II (35 credits)

Answer the questions in only seven of the twelve groups in this part. Be sure to mark the answers to the groups of questions you choose in accordance with the instructions on the front cover of the test booklet. Leave blank the five groups of questions you do not choose to answer.

Group 1 Matter and Energy				
57	1	2	3	4
58	1	2	3	4
59	1	2	3	4
60	1	2	3	4
61	1	2	3	4

Group 2 Atomic Structure				
62	1	2	3	4
63	1	2	3	4
64	1	2	3	4
65	1	2	3	4
66	1	2	3	4

Group 3 Bonding				
67	1	2	3	4
68	1	2	3	4
69	1	2	3	4
70	1	2	3	4
71	1	2	3	4

Group 4 Periodic Table				
72	1	2	3	4
73	1	2	3	4
74	1	2	3	4
75	1	2	3	4
76	1	2	3	4

Group 5 Mathematics of Chemistry				
77	1	2	3	4
78	1	2	3	4
79	1	2	3	4
80	1	2	3	4
81	1	2	3	4

Group 6 Kinetics and Equilibrium				
82	1	2	3	4
83	1	2	3	4
84	1	2	3	4
85	1	2	3	4
86	1	2	3	4

Group 7 Acids and Bases				
87	1	2	3	4
88	1	2	3	4
89	1	2	3	4
90	1	2	3	4
91	1	2	3	4

Group 8 Redox and Electrochemistry				
92	1	2	3	4
93	1	2	3	4
94	1	2	3	4
95	1	2	3	4
96	1	2	3	4

Group 9 Organic Chemistry				
97	1	2	3	4
98	1	2	3	4
99	1	2	3	4
100	1	2	3	4
101	1	2	3	4

Group 10 Applications of Chemical Principles				
102	1	2	3	4
103	1	2	3	4
104	1	2	3	4
105	1	2	3	4
106	1	2	3	4

Group 11 Nuclear Chemistry				
107	1	2	3	4
108	1	2	3	4
109	1	2	3	4
110	1	2	3	4
111	1	2	3	4

Group 12 Laboratory Activities				
112	1	2	3	4
113	1	2	3	4
114	1	2	3	4
115	1	2	3	4
116	1	2	3	4

I do hereby affirm, at the close of this examination, that I had no unlawful knowledge of the questions or answers prior to the examination and that I have neither given nor received assistance in answering any of the questions during the examination.

_____

Signature

[OVER]

The University of the State of New York

REGENTS HIGH SCHOOL EXAMINATION

CHEMISTRY

ANSWER SHEET

☐ Male

Student .... Sex: ☐ Female

Teacher ....

School ....

Record all of your answers on this answer sheet in accordance with the instructions on the front cover of the test booklet.

Part I (65 credits)

1	1 2 3 4		21	1 2 3 4		41	1 2 3 4
2	1 2 3 4		22	1 2 3 4		42	1 2 3 4
3	1 2 3 4		23	1 2 3 4		43	1 2 3 4
4	1 2 3 4		24	1 2 3 4		44	1 2 3 4
5	1 2 3 4		25	1 2 3 4		45	1 2 3 4
6	1 2 3 4		26	1 2 3 4		46	1 2 3 4
7	1 2 3 4		27	1 2 3 4		47	1 2 3 4
8	1 2 3 4		28	1 2 3 4		48	1 2 3 4
9	1 2 3 4		29	1 2 3 4		49	1 2 3 4
10	1 2 3 4		30	1 2 3 4		50	1 2 3 4
11	1 2 3 4		31	1 2 3 4		51	1 2 3 4
12	1 2 3 4		32	1 2 3 4		52	1 2 3 4
13	1 2 3 4		33	1 2 3 4		53	1 2 3
14	1 2 3 4		34	1 2 3 4		54	1 2 3
15	1 2 3 4		35	1 2 3 4		55	1 2 3
16	1 2 3 4		36	1 2 3 4		56	1 2 3
17	1 2 3 4		37	1 2 3 4		
18	1 2 3 4		38	1 2 3 4		
19	1 2 3 4		39	1 2 3 4		
20	1 2 3 4		40	1 2 3 4		

Your answers for Part II should be placed in the proper spaces on the back of this sheet.

Part II (35 credits)

Answer the questions in only seven of the twelve groups in this part. Be sure to mark the answers to the groups of questions you choose in accordance with the instructions on the front cover of the test booklet. Leave blank the five groups of questions you do not choose to answer.

Group 1 Matter and Energy				
57	1	2	3	4
58	1	2	3	4
59	1	2	3	4
60	1	2	3	4
61	1	2	3	4

Group 2 Atomic Structure				
62	1	2	3	4
63	1	2	3	4
64	1	2	3	4
65	1	2	3	4
66	1	2	3	4

Group 3 Bonding				
67	1	2	3	4
68	1	2	3	4
69	1	2	3	4
70	1	2	3	4
71	1	2	3	4

Group 4 Periodic Table				
72	1	2	3	4
73	1	2	3	4
74	1	2	3	4
75	1	2	3	4
76	1	2	3	4

Group 5 Mathematics of Chemistry				
77	1	2	3	4
78	1	2	3	4
79	1	2	3	4
80	1	2	3	4
81	1	2	3	4

Group 6 Kinetics and Equilibrium				
82	1	2	3	4
83	1	2	3	4
84	1	2	3	4
85	1	2	3	4
86	1	2	3	4

Group 7 Acids and Bases				
87	1	2	3	4
88	1	2	3	4
89	1	2	3	4
90	1	2	3	4
91	1	2	3	4

Group 8 Redox and Electrochemistry				
92	1	2	3	4
93	1	2	3	4
94	1	2	3	4
95	1	2	3	4
96	1	2	3	4

Group 9 Organic Chemistry				
97	1	2	3	4
98	1	2	3	4
99	1	2	3	4
100	1	2	3	4
101	1	2	3	4

Group 10 Applications of Chemical Principles				
102	1	2	3	4
103	1	2	3	4
104	1	2	3	4
105	1	2	3	4
106	1	2	3	4

Group 11 Nuclear Chemistry				
107	1	2	3	4
108	1	2	3	4
109	1	2	3	4
110	1	2	3	4
111	1	2	3	4

Group 12 Laboratory Activities				
112	1	2	3	4
113	1	2	3	4
114	1	2	3	4
115	1	2	3	4
116	1	2	3	4

I do hereby affirm, at the close of this examination, that I had no unlawful knowledge of the questions or answers prior to the examination and that I have neither given nor received assistance in answering any of the questions during the examination.

_____
Signature

[OVER]

Free!

Did you know that The Microsoft Network gives you one free month?

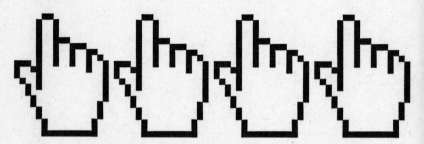

Call us at 1-800-FREE MSN. We'll send you a free CD to get you going.

Then, you can explore the World Wide Web for one month, free. Exchange e-mail with your family and friends. Play games, book airline tickets, handle finances, go car shopping, explore old hobbies and discover new ones. There's one big, useful online world out there. And for one month, it's a free world.

Call **1-800-FREE MSN,** Dept. 3197, for offer details or visit us at **www.msn.com**. Some restrictions apply.

Microsoft Where do you want to go today?®

MSn.
The Microsoft Network

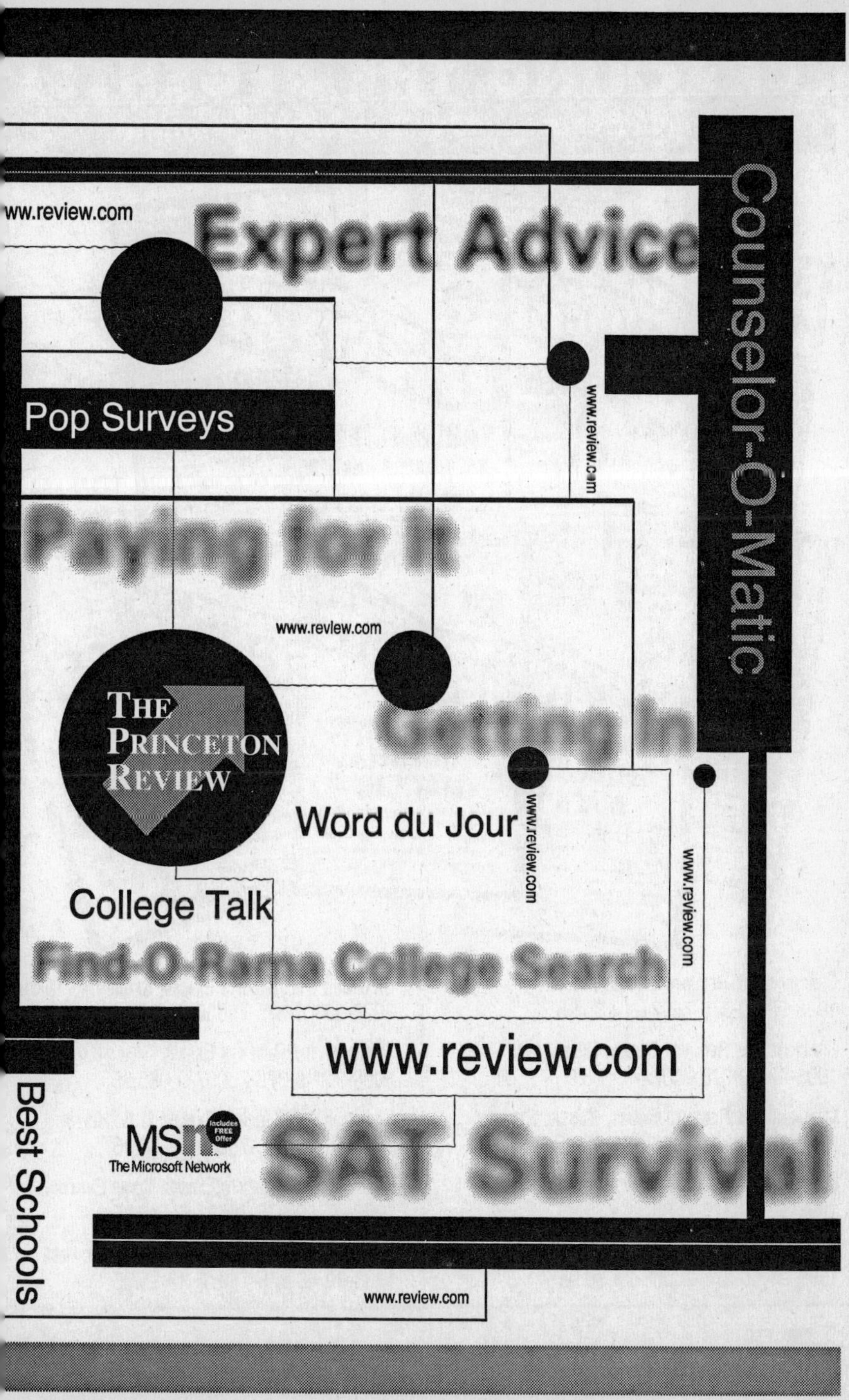
ww.review.com
Expert Advice
Counselor-O-Matic
Pop Surveys
www.review.com
Paying for It
www.review.com
The Princeton Review
Getting In
Word du Jour
www.review.com
www.review.com
College Talk
Find-O-Rama College Search
www.review.com
Best Schools
MSN The Microsoft Network
Includes FREE Offer
SAT Survival
www.review.com

CRACKING THE REGENTS!

Practice exams and test-cracking techniques